RISE OF THE ROBOTS

ALSO BY Martin Ford:

The Lights in the Tunnel: Automation, Accelerating Technology and the Economy of the Future

Praise for *Rise of the Robots*

Longlisted for the 2015 800-CEO-READ Business Book Award

"This is both a humbling book and, in the best sense, a humble book. . . . As Martin Ford documents in *Rise of the Robots*, the job-eating maw of technology now threatens even the nimblest and most expensively educated . . . the human consequences of robotization are already upon us, and skillfully chronicled here. . . . Ford, a software entrepreneur who both understands the technology and has made a thorough study of its economic consequences, never succumbs to the obvious temptation to overdramatize or exaggerate...as a first step toward a solution, Ford's may be the best that the feeble human mind can come up with at the moment." —*New York Times Book Review*

"Mr. Ford lucidly sets out myriad examples of how focused applications of versatile machines (coupled with human helpers where necessary) could displace or de-skill many jobs. . . . You will likely read with mounting dismay Mr. Ford's compelling explanation of how tools that encapsulate 'analytic intelligence and institutional knowledge' will enable less-qualified rivals to carry out your job proficiently, quite possibly from another country." —*Wall Street Journal*

"*Rise of the Robots* is about as scary as the title suggests. It's not science fiction, but rather a vision (almost) of economic Armageddon."

—Frank Bruni, *New York Times*

"Compelling and well-written. . . . In his conception, the answer is a combination of short-term policies and longer-term initiatives, one of which is a radical idea that may gain some purchase among gloomier techno-profits: a guaranteed income for all citizens. If that stirs up controversy, that's the point. The book is both lucid and bold, and certainly a starting point for robust debate about the future of all workers in an age of advancing robotics and looming artificial intelligence systems."

—ZDNet

"An alarming new book." —*Esquire*

"A thorough look at how far machines have come."
—*Washington Post*, Innovations blog

"*Rise of the Robots* is an excellent book. Fair-minded, balanced, well-researched, and fully thought through."
—Inside Higher Ed, Learn blog

"[Ford's] a careful and thoughtful writer who relies on ample evidence, clear reasoning, and lucid economic analysis. In other words, it's entirely possible that he's right." —Daily Beast

"In *Rise of the Robots*, Ford coolly and clearly considers what work is under threat from automation." —*New Scientist*

"Surveying all the fields now being affected by automation, Ford makes a compelling case that this is an historic disruption—a fundamental shift from most tasks being performed by humans to one where most tasks are done by machines." —*Fast Company*

"Makes clear the need to come to grips with ever more rapidly advancing technology and its effects on how people make a living and how the economy functions." —*Pittsburgh Tribune-Review*

"Ford offers ideas on changes in social policies, including guaranteed income, to keep our economy humming and prepare ourselves for a more automated future." —*Booklist*

"A careful and courageous examination of automation and its possible impact on society." —*Kirkus Reviews*

"Well-written with interesting stories about both business and technology." —*Wired*, Dot Physics blog

"If *The Second Machine Age* was 2014's tech-economy title of choice, this book may be 2015's equivalent. Ford has a far more pessimistic take, persuasively arguing that few roles will escape the current wave of technological disruption." —Andrew Hill, *Financial Times*

"Robots, and their like, are on the rise. Their impact will be an important question in the next decade and beyond. Martin Ford has been thinking in this area before most others, so this book deserves very careful consideration."

—Lawrence Summers, president emeritus and Charles W. Eliot University Professor, Harvard University

"Martin Ford's *Rise of the Robots* is a very important, timely, and well-informed book. Smart machines, artificial intelligence, nanotechnology, and the Internet of things are transforming every sector of the economy. Machines can outperform workers in a rapidly widening arc of activities. Will smart machines lead to a world of plenty, leisure, health care, and education for all; or to a world of inequality, mass unemployment, and a war between the haves and have-nots, and between the machines and the workers left behind? Ford doesn't claim to have all of the answers, but he asks the right questions and offers a highly informed and panoramic view of the debate. This is an excellent book that offers us a sophisticated glimpse into our possible futures."

—Jeffrey D. Sachs, director, Earth Institute, Columbia University, and author of *The Age of Sustainable Development*

"Ever since the Luddites, pessimists have believed that technology would destroy jobs. So far they have been wrong. Martin Ford shows with great clarity why today's automated technology will be much more destructive of jobs than previous technological innovation. This is a book that everyone concerned with the future of work must read."

—Lord Robert Skidelsky, emeritus professor of political economy, University of Warwick, and author of a three-volume biography of John Maynard Keynes

"Martin Ford has thrust himself into the center of the debate over AI, big data, and the future of the economy with a shrewd look at the forces shaping our lives and work. As an entrepreneur pioneering many of the trends he uncovers, he speaks with special credibility, insight, and verve. Business people, policy makers, and professionals of all sorts should read this book right away—before the 'bots steal their jobs. Ford gives us a roadmap to the future."

—Kenneth Cukier, data editor, *Economist*,
and co-author of *Big Data: A Revolution That Will Transform How We Live, Work, and Think*

"It's not easy to accept, but it's true. Education and hard work will no longer guarantee success for huge numbers of people as technology advances. The time for denial is over. Now it's time to consider solutions and there are very few proposals on the table. *Rise of the Robots* presents one idea, the basic income model, with clarity and force. No one who cares about the future of human dignity can afford to skip this book."

—Jaron Lanier, author of *You Are Not a Gadget*
and *Who Owns the Future?*

"If the robots are coming for my job (too), then Martin Ford is the person I want on my side, not to fend them off but to construct a better world where we can all—humans and our machines—live more prosperously together. *Rise of the Robots* goes far beyond the usual fear-mongering punditry to suggest an action plan for a better future."

—Cathy N. Davidson, distinguished professor and director,
The Futures Initiative, The Graduate Center, CUNY,
and author of *Now You See It: How the Brain Science of Attention Will Transform the Way We Live, Work, and Learn*

RISE OF THE ROBOTS

TECHNOLOGY AND THE THREAT OF A JOBLESS FUTURE

MARTIN FORD

BASIC
BOOKS
New York

Published by Basic Books,
an imprint of Perseus Books LLC.,
a subsidiary of Hachette Book Group, Inc.

First paperback edition published in 2016 by Basic Books.

Books published by Basic Books are available at special discounts for bulk purchases in the United States by corporations, institutions, and other organizations. For more information, please contact the Special Markets Department at Perseus Books, 2300 Chestnut Street, Suite 200, Philadelphia, PA 19103, or call (800) 810-4145, ext. 5000, or email special.markets@perseusbooks.com.

Designed by Pauline Brown

Library of Congress Cataloging-in-Publication Data

Ford, Martin (Martin R.)
Rise of the robots : technology and the threat of a jobless future / Martin Ford.
pages cm
Includes bibliographical references and index.
ISBN 978-0-465-05999-7 (hardback) — ISBN 978-0-465-04067-4 (e-book) 1. Labor supply—Effect of automation on. 2. Labor supply—Effect of technological innovations on. 3. Employment forecasting. 4. Technological innovations—Economic aspects. I. Title.
HD6331.F58 2015
331.13'7042—dc23

2014041327

ISBN 978-0-465-09753-1 (paperback)

LSC-C

10

To Tristan, Colin,
Elaine, and Xiaoxiao

CONTENTS

INTRODUCTION

Sometime during the 1960s, the Nobel laureate economist Milton Friedman was consulting with the government of a developing Asian nation. Friedman was taken to a large-scale public works project, where he was surprised to see large numbers of workers wielding shovels, but very few bulldozers, tractors, or other heavy earth-moving equipment. When asked about this, the government official in charge explained that the project was intended as a "jobs program." Friedman's caustic reply has become famous: "So then, why not give the workers spoons instead of shovels?"

Friedman's remark captures the skepticism—and often outright derision—expressed by economists confronting fears about the prospect of machines destroying jobs and creating long-term unemployment. Historically, that skepticism appears to be well-founded. In the United States, especially during the twentieth century, advancing technology has consistently driven us toward a more prosperous society.

There have certainly been hiccups—and indeed major disruptions—along the way. The mechanization of agriculture vaporized millions of jobs and drove crowds of unemployed farmhands into cities in search of factory work. Later, automation and globalization pushed workers out of the manufacturing sector and into new service jobs. Short-term unemployment was often a problem during these transitions, but it never became systemic or permanent. New jobs were created and dispossessed workers found new opportunities.

What's more, those new jobs were often better than earlier counterparts, requiring upgraded skills and offering better wages. At no time was this more true than in the two and a half decades following World War II. This "golden age" of the American economy was characterized by a seemingly perfect symbiosis between rapid technological progress and the welfare of the American workforce. As the machines used in production improved, the productivity of the workers operating those machines likewise increased, making them more valuable and allowing them to demand higher wages. Throughout the postwar period, advancing technology deposited money directly into the pockets of average workers as their wages rose in tandem with soaring productivity. Those workers, in turn, went out and spent their ever-increasing incomes, further driving demand for the products and services they were producing.

As that virtuous feedback loop powered the American economy forward, the profession of economics was enjoying its own golden age. It was during the same period that towering figures like Paul Samuelson worked to transform economics into a science with a strong mathematical foundation. Economics gradually came to be almost completely dominated by sophisticated quantitative and statistical techniques, and economists began to build the complex mathematical models that still constitute the field's intellectual basis. As the postwar economists did their work, it would have been natural for them to look at the thriving economy around them and assume that it was normal: that it was the way an economy was supposed to work—*and would always work*.

In his 2005 book *Collapse: How Societies Choose to Succeed or Fail,* Jared Diamond tells the story of agriculture in Australia. In the nineteenth century, when Europeans first colonized Australia, they found a relatively lush, green landscape. Like American economists in the 1950s, the Australian settlers assumed that what they were seeing was normal, and that the conditions they observed would continue indefinitely. They invested heavily in developing farms and ranches on this seemingly fertile land.

Within a decade or two, however, reality struck. The farmers found that the overall climate was actually far more arid than they were initially led to believe. They had simply had the good fortune (or perhaps misfortune) to arrive during a climactic "Goldilocks period"—a sweet spot when everything happened to be just right for agriculture. Today in Australia, you can find the remnants of those ill-fated early investments: abandoned farm houses in the middle of what is essentially a desert.

There are good reasons to believe that America's economic Goldilocks period has likewise come to an end. That symbiotic relationship between increasing productivity and rising wages began to dissolve in the 1970s. As of 2013, a typical production or nonsupervisory worker earned about 13 percent less than in 1973 (after adjusting for inflation), even as productivity rose by 107 percent and the costs of big-ticket items like housing, education, and health care have soared.[1]

On January 2, 2010, the *Washington Post* reported that the first decade of the twenty-first century resulted in the creation of no new jobs. Zero.[2] This hasn't been true of any decade since the Great Depression; indeed, there has never been a postwar decade that produced less than a 20 percent increase in the number of available jobs. Even the 1970s, a decade associated with stagflation and an energy crisis, generated a 27 percent increase in jobs.[3] The lost decade of the 2000s is especially astonishing when you consider that the US economy needs to create roughly a million jobs per year just to keep up with growth in the size of the workforce. In other words, during those first ten years there were about 10 million missing jobs that should have been created—but never showed up.

Income inequality has since soared to levels not seen since 1929, and it has become clear that the productivity increases that went into workers' pockets back in the 1950s are now being retained almost entirely by business owners and investors. The share of overall national income going to labor, as opposed to capital, has fallen precipitously and appears to be in continuing free fall. Our Goldilocks period has reached its end, and the American economy is moving into a new era.

It is an era that will be defined by a fundamental shift in the relationship between workers and machines. That shift will ultimately challenge one of our most basic assumptions about technology: that *machines are tools* that increase the productivity of workers. Instead, machines themselves are turning into workers, and the line between the capability of labor and capital is blurring as never before.

All this progress is, of course, being driven by the relentless acceleration in computer technology. While most people are by now familiar with Moore's Law—the well-established rule of thumb that says computing power roughly doubles every eighteen to twenty-four months—not everyone has fully assimilated the implications of this extraordinary exponential progress.

Imagine that you get in your car and begin driving at 5 miles per hour. You drive for a minute, accelerate to double your speed to 10 mph, drive for another minute, double your speed again, and so on. The really remarkable thing is not simply the fact of the doubling but the amount of ground you cover after the process has gone on for a while. In the first minute, you would travel about 440 feet. In the third minute at 20 mph, you'd cover 1,760 feet. In the fifth minute, speeding along at 80 mph, you would go well over a mile. To complete the sixth minute, you'd need a faster car—as well as a racetrack.

Now think about how fast you would be traveling—and how much progress you would make in that final minute—if you doubled your speed twenty-seven times. That's roughly the number of times computing power has doubled since the invention of the integrated circuit in 1958. The revolution now under way is happening not just because of the acceleration itself but because *that acceleration has been going on for so long* that the amount of progress we can now expect in any given year is potentially mind-boggling.

The answer to the question about your speed in the car, by the way, is 671 *million* miles per hour. In that final, twenty-eighth minute, you would travel more than 11 million miles. Five minutes or so at that speed would get you to Mars. That, in a nutshell, is where information technology stands today, relative to when the first primitive integrated circuits started plodding along in the late 1950s.

As someone who has worked in software development for more than twenty-five years, I've had a front-row seat when it comes to observing that extraordinary acceleration in computing power. I've also seen at close hand the tremendous progress made in software design, and in the tools that make programmers more productive. And, as a small business owner, I've watched as technology has transformed the way I run my business—in particular, how it has dramatically reduced the need to hire employees to perform many of the routine tasks that have always been essential to the operation of any business.

In 2008, as the global financial crisis unfolded, I began to give serious thought to the implications of that consistent doubling in computational power and, especially, to the likelihood that it would dramatically transform the job market and overall economy in coming years and decades. The result was my first book, *The Lights in the Tunnel: Automation, Accelerating Technology and the Economy of the Future*, published in 2009.

In that book, even as I wrote about the importance of accelerating technology, I underestimated just how rapidly things would in fact move forward. For example, I noted that auto manufacturers were working on collision avoidance systems to help prevent accidents, and I suggested that "over time these systems could evolve into technology capable of driving the car autonomously." Well, it turned out that "over time" wasn't much time at all! Within a year of the book's publication, Google introduced a fully automated car capable of driving in traffic. And since then, three states—Nevada, California, and Florida—have passed laws allowing self-driving vehicles to share the road on a limited basis.

I also wrote about progress being made in the field of artificial intelligence. At the time, the story of IBM's "Deep Blue" computer and how it had defeated world chess champion Garry Kasparov in 1997, was perhaps the most impressive demonstration of AI in action. Once again, I was taken by surprise when IBM introduced Deep Blue's successor, Watson—a machine that took on a far more difficult challenge: the television game show *Jeopardy!* Chess is a game with rigidly defined rules; it is the sort of thing we might expect a computer to be good at. *Jeopardy!* is something else entirely: a game that draws on an almost limitless body of knowledge and requires a sophisticated ability to parse language, including even jokes and puns. Watson's success at *Jeopardy!* is not only impressive, it is highly practical, and in fact, IBM is already positioning Watson to play a significant role in fields like medicine and customer service.

It's a good bet that nearly all of us will be surprised by the progress that occurs in the coming years and decades. Those surprises won't be confined to the nature of the technical advances themselves: the impact that accelerating progress has on the job market and the overall economy is poised to defy much of the conventional wisdom about how technology and economics intertwine.

One widely held belief that is certain to be challenged is the assumption that automation is primarily a threat to workers who have little education and lower-skill levels. That assumption emerges from the fact that such jobs tend to be routine and repetitive. Before you get too comfortable with that idea, however, consider just how fast the frontier is moving. At one time, a "routine" occupation would probably have implied standing on an assembly line. The reality today is far different. While lower-skill occupations will no doubt continue to be affected, a great many college-educated, white-collar workers are going to discover that their jobs, too, are squarely in the sights as software automation and predictive algorithms advance rapidly in capability.

The fact is that "routine" may not be the best word to describe the jobs most likely to be threatened by technology. A more accurate term might be "predictable." Could another person learn to do your job by studying a detailed record of everything you've done in the past? Or could someone become proficient by repeating the tasks you've already completed, in the way that a student might take practice tests to prepare for an exam? If so, then there's a good chance that an algorithm may someday be able to learn to do much, or all, of your job. That's made especially likely as the "big data" phenomenon continues to unfold: organizations are collecting incomprehensible amounts of information about nearly every aspect of their operations, and a great many jobs and tasks are likely to be encapsulated in that data—waiting for the day when a smart machine learning algorithm comes along and begins schooling itself by delving into the record left by its human predecessors.

The upshot of all this is that acquiring more education and skills will not necessarily offer effective protection against job automation in the future. As an example, consider radiologists, medical doctors who specialize in the interpretation of medical images. Radiologists require a tremendous amount of training, typically a minimum of thirteen years beyond high school. Yet, computers are rapidly getting better at analyzing images. It's quite easy to imagine that someday, in the not too distant future, radiology will be a job performed almost exclusively by machines.

In general, computers are becoming very proficient at acquiring skills, especially when a large amount of training data is available. Entry-level jobs, in particular, are likely to be heavily affected, and there is evidence that this may already be occurring. Wages for new college graduates have actually been declining over the past decade, while up to 50 percent of new graduates are forced to take jobs that do not require a college degree. Indeed, as I'll demonstrate in this book, employment for many skilled professionals—including lawyers, journalists, scientists, and pharmacists—is already being significantly eroded by advancing information technology. They are not alone: most jobs are, on some level, fundamentally routine and predictable, with relatively few people paid primarily to engage in truly creative work or "blue-sky" thinking.

As machines take on that routine, predictable work, workers will face an unprecedented challenge as they attempt to adapt. In the past, automation technology has tended to be relatively specialized and to disrupt one employment sector at a time, with workers then switching to a new emerging industry. The situation today is quite different. Information technology is a truly general-purpose technology, and its impact will occur across the board. Virtually every industry in existence is likely to become less labor-intensive as new technology is assimilated into business models—and that transition could happen quite rapidly. At the same time, the new industries that emerge will nearly always incorporate powerful labor-saving technology right from their inception. Companies like Google and Facebook, for example, have succeeded in becoming household names and achieving massive market valuations while hiring only a tiny number of people relative to their size and influence. There's every reason to expect that a similar scenario will play out with respect to nearly all the new industries created in the future.

All of this suggests that we are headed toward a transition that will put enormous stress on both the economy and society. Much of the conventional advice offered to workers and to students who are preparing to enter the workforce is likely to be ineffective. The unfortunate reality is that a great many people will do everything right—at least in terms of pursuing higher education and acquiring skills—and yet will still fail to find a solid foothold in the new economy.

Beyond the potentially devastating impact of long-term unemployment and underemployment on individual lives and on the fabric of society, there will also be a significant economic price. The virtuous feedback loop between productivity, rising wages, and increasing consumer spending will collapse. That positive feedback effect is already seriously diminished: we face soaring inequality not just in income but also in consumption. The top 5 percent of households are currently responsible for nearly 40 percent of spending, and that trend toward increased concentration at the top seems almost certain to continue. Jobs remain the primary mechanism by which purchasing power gets into the hands of consumers. If that mechanism continues to erode, we will face the prospect of having too few viable consumers to continue driving economic growth in our mass-market economy.

As this book will make clear, advancing information technology is pushing us toward a tipping point that is poised to ultimately make the entire economy less labor-intensive. However, that transition won't necessarily unfold in a uniform or predictable way. Two sectors in particular—higher education and health care—have, so far, been highly resistant to the kind of disruption that is already becoming evident in the broader economy. The irony is that the failure of technology to transform these sectors could amplify its negative consequences elsewhere, as the costs of health care and education become ever more burdensome.

Technology, of course, will not shape the future in isolation. Rather, it will intertwine with other major societal and environmental challenges such as an aging population, climate change, and resource depletion. It's often predicted that a shortage of workers will eventually develop as the baby boom generation exits the workforce, effectively counterbalancing—or perhaps even overwhelming—any impact from automation. Rapid innovation is typically framed purely as a countervailing force with the potential to minimize, or even reverse, the stress we put on the environment. However, as we'll see, many of these assumptions rest on uncertain foundations: the story is sure to be far more complicated. Indeed, the frightening reality is that if we don't recognize and adapt to the implications of advancing technology, we may face the prospect of a "perfect storm" where the impacts from soaring inequality, technological unemployment, and climate change unfold roughly in parallel, and in some ways amplify and reinforce each other.

In Silicon Valley the phrase "disruptive technology" is tossed around on a casual basis. No one doubts that technology has the power to devastate entire industries and upend specific sectors of the economy and job market. The question I will ask in this book is bigger: Can accelerating technology disrupt *our entire system* to the point where a fundamental restructuring may be required if prosperity is to continue?

Chapter 1

THE AUTOMATION WAVE

A warehouse worker approaches a stack of boxes. The boxes are of varying shapes, sizes, and colors, and they are stacked in a somewhat haphazard way.

Imagine for a moment that you can see inside the brain of the worker tasked with moving the boxes, and consider the complexity of the problem that needs to be solved.

Many of the boxes are a standard brown color and are pressed tightly against each other, making the edges difficult to perceive. Where precisely does one box end and the next begin? In other cases, there are gaps and misalignments. Some boxes are rotated so that one edge juts out. At the top of the pile, a small box rests at an angle in the space between two larger boxes. Most of the boxes are plain brown or white cardboard, but some are emblazoned with company logos, and a few are full-color retail boxes intended to be displayed on store shelves.

The human brain is, of course, capable of making sense of all this complicated visual information almost instantaneously. The worker easily perceives the dimensions and orientation of each box, and seems to know instinctively that he must begin by moving the boxes at the top of the stack and how to move the boxes in a sequence that won't destabilize the rest of the pile.

This is exactly the type of visual perception challenge that the human brain has evolved to overcome. That the worker succeeds in moving the boxes would be completely unremarkable—were it not for the fact that, in this case, the worker is a robot. To be more precise, it is a snake-like robotic arm, its head consisting of a suction-powered gripper. The robot is slower to comprehend than a human would be. It peers at the boxes, adjusts its gaze slightly, ponders some more, and then finally lunges forward and grapples a box from the top of the pile.* The sluggishness, however, results almost entirely from the staggering complexity of the computation required to perform this seemingly simple task. If there is one thing the history of information technology teaches, it is that this robot is going to very soon get a major speed upgrade.

Indeed, engineers at Industrial Perception, Inc., the Silicon Valley start-up company that designed and built the robot, believe the machine will ultimately be able to move a box every second. That compares with a human worker's maximum rate of a box roughly every six seconds.[1] Needless to say, the robot can work continuously; it will never get tired or suffer a back injury—and it will certainly never file a worker's compensation claim.

Industrial Perception's robot is remarkable because its capability sits at the nexus of visual perception, spatial computation, and dexterity. In other words, it is invading the final frontier of machine automation, where it will compete for the few relatively routine, manual jobs that are still available to human workers.

Robots in factories are, of course, nothing new. They have become indispensable in virtually every sector of manufacturing, from

* A video of Industrial Perception's box-moving robot can be seen on the company's website at http://www.industrial-perception.com/technology.html.

automobiles to semiconductors. Electric-car company Tesla's new plant in Fremont, California, uses 160 highly flexible industrial robots to assemble about 400 cars per week. As a new-car chassis arrives at the next position in the assembly line, multiple robots descend on it and operate in coordination. The machines are able to autonomously swap the tools wielded by their robotic arms in order to complete a variety of tasks. The same robot, for example, installs the seats, retools itself, and then applies adhesive and drops the windshield into place.[2] According to the International Federation of Robotics, global shipments of industrial robots increased by more than 60 percent between 2000 and 2012, with total sales of about $28 billion in 2012. By far the fastest-growing market is China, where robot installations grew at about 25 percent per year between 2005 and 2012.[3]

While industrial robots offer an unrivaled combination of speed, precision, and brute strength, they are, for the most part, blind actors in a tightly choreographed performance. They rely primarily on precise timing and positioning. In the minority of cases where robots have machine vision capability, they can typically see in just two dimensions and only in controlled lighting conditions. They might, for example, be able to select parts from a flat surface, but an inability to perceive depth in their field of view results in a low tolerance for environments that are to any meaningful degree unpredictable. The result is that a number of routine factory jobs have been left for people. Very often these are jobs that involve filling the gaps between the machines, or they are at the end points of the production process. Examples might include choosing parts from a bin and then feeding them into the next machine, or loading and unloading the trucks that move products to and from the factory.

The technology that powers the Industrial Perception robot's ability to see in three dimensions offers a case study in the ways that cross-fertilization can drive bursts of innovation in unexpected areas. It might be argued that the robot's eyes can trace their origin to November 2006, when Nintendo introduced its Wii video game console.

Nintendo's machine included an entirely new type of game controller: a wireless wand that incorporated an inexpensive device called an accelerometer. The accelerometer was able to detect motion in three dimensions and then output a data stream that could be interpreted by the game console. Video games could now be controlled through body movements and gestures. The result was a dramatically different game experience. Nintendo's innovation smashed the stereotype of the nerdy kid glued to a monitor and a joystick, and opened a new frontier for games as active exercise.

It also demanded a competitive response from the other major players in the video game industry. Sony Corporation, makers of the PlayStation, elected to essentially copy Nintendo's design and introduced its own motion-detecting wand. Microsoft, however, aimed to leapfrog Nintendo and come up with something entirely new. The Kinect add-on to the Xbox 360 game console eliminated the need for a controller wand entirely. To accomplish this, Microsoft built a webcam-like device that incorporates three-dimensional machine vision capability based in part on imaging technology created at a small Israeli company called PrimeSense. The Kinect sees in three dimensions by using what is, in essence, sonar at the speed of light: it shoots an infrared beam at the people and objects in a room and then calculates their distance by measuring the time required for the reflected light to reach its infrared sensor. Players could now interact with the Xbox game console simply by gesturing and moving in view of the Kinect's camera.

The truly revolutionary thing about the Kinect was its price. Sophisticated machine vision technology—which might previously have cost tens or even hundreds of thousands of dollars and required bulky equipment—was now available in a compact and lightweight consumer device priced at $150. Researchers working in robotics instantly realized the potential for the Kinect technology to transform their field. Within weeks of the product's introduction, both university-based engineering teams and do-it-yourself innovators had hacked into the Kinect and posted YouTube videos of robots that were now able to see in three dimensions.[4] Industrial Perception likewise decided to base its vision system on the technology that powers the Kinect, and the result is an affordable machine that is rapidly approaching a nearly human-level ability to perceive and interact with its environment while dealing with the kind of uncertainty that characterizes the real world.

A Versatile Robotic Worker

Industrial Perception's robot is a highly specialized machine focused specifically on moving boxes with maximum efficiency. Boston-based Rethink Robotics has taken a different track with Baxter, a lightweight humanoid manufacturing robot that can easily be trained to perform a variety of repetitive tasks. Rethink was founded by Rodney Brooks, one of the world's foremost robotics researchers at MIT and a co-founder of iRobot, the company that makes the Roomba automated vacuum cleaner as well as military robots used to defuse bombs in Iraq and Afghanistan. Baxter, which costs significantly less than a year's wages for a typical US manufacturing worker, is essentially a scaled-down industrial robot that is designed to operate safely in close proximity to people.

In contrast to industrial robots, which require complex and expensive programming, Baxter can be trained simply by moving its arms through the required motions. If a facility uses multiple robots, one Baxter can be trained and then the knowledge can be propagated to the others simply by plugging in a USB device. The robot can be adapted to a variety of tasks, including light assembly work, transferring parts between conveyer belts, packing products into retail packaging, or tending machines used in metal fabrication. Baxter is particularly talented at packing finished products into shipping boxes. K'NEX, a toy construction set manufacturer located in Hatfield, Pennsylvania, found that Baxter's ability to pack its products tightly allowed the company to use 20–40 percent fewer boxes.[5] Rethink's robot also has two-dimensional machine vision capability powered by cameras on both wrists and can pick up parts and even perform basic quality-control inspections.

The Coming Explosion in Robotics

While Baxter and Industrial Perception's box-moving robot are dramatically different machines, they are both built on the same fundamental software platform. ROS—or Robot Operating System—was originally conceived at Stanford University's Artificial Intelligence Laboratory and then developed into a full-fledged robotics platform by Willow Garage, Inc., a small company that designs and manufactures programmable robots that are used primarily by researchers at universities. ROS is similar to operating systems like Microsoft Windows, Macintosh OS, or Google's Android but is geared specifically toward making robots easy to program and control. Because ROS is free and also open source—meaning that software developers can easily modify and enhance it—it is rapidly becoming the standard software platform for robotics development.

The history of computing shows pretty clearly that once a standard operating system, together with inexpensive and easy-to-use programming tools, becomes available, an explosion of application software is likely to follow. This has been the case with personal computer software and, more recently, with iPhone, iPad, and Android apps. Indeed, these platforms are now so saturated with application software that it can be genuinely difficult to conceive of an idea that hasn't already been implemented.

It's a good bet that the field of robotics is poised to follow a similar path; we are, in all likelihood, at the leading edge of an explosive wave of innovation that will ultimately produce robots geared toward nearly every conceivable commercial, industrial, and consumer task. That explosion will be powered by the availability of standardized software and hardware building blocks that will make it a relatively simple matter to assemble new designs without the need to reinvent the wheel. Just as the Kinect made machine vision affordable, other hardware components—such as robotic arms—will see their costs driven down as robots begin scaling up to high-volume production. As of 2013, there were already thousands of software components available to work with ROS, and development platforms were cheap enough to allow nearly anyone to start designing new robotics applications. Willow Garage, for example, sells a complete mobile robot kit called TurtleBot that includes Kinect-powered machine vision for about $1,200. After inflation is taken into account, that's far less than what an inexpensive personal computer and monitor cost in the early 1990s, when Microsoft Windows was in the early stages of producing its own software explosion.

When I visited the RoboBusiness conference and tradeshow in Santa Clara, California, in October 2013, it was clear that the robotics industry had already started gearing up for the coming explosion. Companies of all sizes were on hand to showcase robots designed to perform precision manufacturing, transport medical supplies between departments in large hospitals, or autonomously operate heavy equipment for agriculture and mining. There was a personal robot named "Budgee" capable of carrying up to fifty pounds of stuff around the house or at the store. A variety of educational robots focused on everything from encouraging technical creativity to assisting children with autism or learning disabilities. At the Rethink Robotics booth, Baxter had received Halloween training and was grasping small boxes of candy and then dropping them into pumpkin-shaped trick-or-treat buckets. There were also companies marketing components like motors, sensors, vision systems, electronic controllers, and the specialized software used to construct robots. Silicon Valley start-up Grabit Inc. demonstrated an innovative electroadhesion-powered gripper that allows robots to pick up, carry, and place nearly anything simply by employing a controlled electrostatic charge. To round things out, a global law firm with a specialized robotics practice was on hand to help employers navigate the complexities of labor, employment, and safety regulations when robots are brought in to replace, or work in close proximity to, people.

One of the most remarkable sights at the tradeshow was in the aisles—which were populated by a mix of human attendees and dozens of remote-presence robots provided by Suitable Technologies, Inc. These robots, consisting of a flat screen and camera mounted on a mobile pedestal, allowed remote participants to visit tradeshow booths, view demonstrations, ask questions, and otherwise interact normally with other participants. Suitable Technologies offered remote presence at the tradeshow for a minimal fee, allowing visitors from outside the San Francisco Bay area to avoid thousands of dollars in travel costs. After a few minutes, the robots—each with a human face displayed on its screen—did not seem at all out of place as they prowled between booths and engaged other attendees in conversation.

Manufacturing Jobs and Factory Reshoring

In a September 2013 article, Stephanie Clifford of the *New York Times* told the story of Parkdale Mills, a textile factory in Gaffney, South Carolina. The Parkdale plant employs about 140 people. In 1980, the same level of production would have required more than 2,000 factory workers. Within the Parkdale plant, "only infrequently does a person interrupt the automation, mainly because certain tasks are still cheaper if performed by hand—like moving half-finished yarn between machines on forklifts."[6] Completed yarn is conveyed automatically toward packing and shipping machines along pathways attached to the ceiling.

Nonetheless, those 140 factory jobs represent at least a partial reversal of a decades-long decline in manufacturing employment. The

US textile industry was decimated in the 1990s as production moved to low-wage countries, especially China, India, and Mexico. About 1.2 million jobs—more than three-quarters of domestic employment in the textile sector—vanished between 1990 and 2012. The last few years, however, have seen a dramatic rebound in production. Between 2009 and 2012, US textile and apparel exports rose by 37 percent to a total of nearly $23 billion.[7] The turnaround is being driven by automation technology so efficient that it is competitive with even the lowest-wage offshore workers.

Within the manufacturing sector in the United States and other developed countries, the introduction of these sophisticated labor-saving innovations is having a mixed impact on employment. While factories like Parkdale don't directly create large numbers of manufacturing jobs, they do drive increased employment at suppliers and in peripheral areas like driving the trucks that move raw materials and finished products. While a robot like Baxter can certainly eliminate the jobs of some workers who perform routine tasks, it also helps make US manufacturing more competitive with low-wage countries. Indeed, there is now a significant "reshoring" trend under way, and this is being driven both by the availability of new technology and by rising offshore labor costs, especially in China where typical factory workers saw their pay increase by nearly 20 percent per year between 2005 and 2010. In April 2012, the Boston Consulting Group surveyed American manufacturing executives and found that nearly half of companies with sales exceeding $10 billion were either actively pursuing or considering bringing factories back to the United States.[8]

Factory reshoring dramatically decreases transportation costs and also provides many other advantages. Locating factories in close proximity to both consumer markets and product design centers allows companies to cut production lead times and be far more responsive to their customers. As automation becomes ever more flexible and sophisticated, it's likely that manufacturers will trend toward offering more customizable products—perhaps, for example, allowing customers to create unique designs or specify hard-to-find clothing sizes through easy-to-use online interfaces. Domestic automated production could then put a finished product into a customer's hands within days.

There is, however, one important caveat to the reshoring narrative. Even the relatively small number of new factory jobs now being created as a result of reshoring won't necessarily be around over the long term; as robots continue to get more capable and dexterous and as new technologies like 3D printing come into widespread use, it seems likely that many factories will eventually approach full automation. Manufacturing jobs in the United States currently account for well under 10 percent of total employment. As a result, manufacturing robots and reshoring are likely to have a fairly marginal impact on the overall job market.

The story will be very different in developing countries like China, where employment is far more focused in the manufacturing sector. In fact, advancing technology has already had a dramatic impact on Chinese factory jobs; between 1995 and 2002 China lost about 15 percent of its manufacturing workforce, or about 16 million jobs.[9] There is strong evidence to suggest that this trend is poised to accelerate. In 2012, Foxconn—the primary contract manufacturer of Apple devices—announced plans to eventually introduce up to a million robots in its factories. Taiwanese company Delta Electronics, Inc., a producer of power adapters, has recently shifted its strategy to focus on low-cost robots for precision electronics assembly. Delta hopes to offer a one-armed assembly robot for about $10,000—less than half the cost of Rethink's Baxter. European industrial robot manufacturers like ABB Group and Kuka AG are likewise investing heavily in the Chinese market and are currently building local factories to churn out thousands of robots per year.[10]

Increased automation is also likely to be driven by the fact that the interest rates paid by large companies in China are kept artificially low as a result of government policy. Loans are often rolled over continuously, so that the principal is never repaid. This makes capital investment extremely attractive even when labor costs are low and has been one of the primary reasons that investment now accounts for nearly half of China's GDP.[11] Many analysts believe that this artificially low cost of capital has caused a great deal of mal-investment throughout China, perhaps most famously the construction of "ghost cities" that appear to be largely unoccupied. By the same token, low capital costs may create a powerful incentive for big companies to invest in expensive automation, even in those cases where it does not necessarily make good business sense to do so.

One of the biggest challenges for a transition to robotic assembly in the Chinese electronics industry will be designing robots that are flexible enough to keep up with rapid product lifecycles. Foxconn, for example, maintains massive facilities where workers live onsite in dormitories. In order to accommodate aggressive production schedules, thousands of workers can be woken in the middle of the night and set immediately to work. That results in an astonishing ability to rapidly ramp up production or adjust to product design changes, but it also puts extreme pressure on workers—as evidenced by the near epidemic of suicides that occurred at Foxconn facilities in 2010. Robots, of course, have the ability to work continuously, and as they become more flexible and easier to train for new tasks, they will become an increasingly attractive alternative to human workers, even when wages are low.

The trend toward increased factory automation in developing countries is by no means limited to China. Clothing and shoe production, for example, continues to be one of the most labor-intensive sectors of manufacturing, and factories have been transitioning from China to even lower-wage countries like Vietnam and Indonesia. In June 2013, athletic-shoe manufacturer Nike announced that rising wages in Indonesia had negatively impacted its quarterly financial numbers. According to the company's chief financial officer, the long-term solution to that problem is going to be "engineering the labor out of the product."[12] Increased automation is also seen as a way to deflect criticism regarding the sweatshop-like environments that often exist in third-world garment factories.

The Service Sector: Where the Jobs Are

In the United States and other advanced economies, the major disruption will be in the service sector—which is, after all, where the vast majority of workers are now employed. This trend is already evident in areas like ATMs and self-service checkout lanes, but the next decade is likely to see an explosion of new forms of service sector automation, potentially putting millions of relatively low-wage jobs at risk.

San Francisco start-up company Momentum Machines, Inc., has set out to fully automate the production of gourmet-quality hamburgers. Whereas a fast food worker might toss a frozen patty onto the grill, Momentum Machines' device shapes burgers from freshly ground meat and then grills them to order—including even the ability to add just the right amount of char while retaining all the juices. The machine, which is capable of producing about 360 hamburgers per hour, also toasts the bun and then slices and adds fresh ingredients like tomatoes, onions, and pickles only after the order is placed. Burgers arrive assembled and ready to serve on a conveyer belt. While most robotics companies take great care to spin a positive tale when it comes to the potential impact on employment, Momentum Machines co-founder Alexandros Vardakostas is very forthright about the company's objective: "Our device isn't meant to make employees more efficient," he said. "It's meant to completely obviate them."[13] * The company estimates that the average fast food

* The company is not unaware of the potential impact its technology will have on jobs and, according to its website, plans to support a program that will offer discounted technical training to workers who are displaced.

restaurant spends about $135,000 per year on wages for employees who produce hamburgers and that the total labor cost for burger production for the US economy is about $9 billion annually.[14] Momentum Machines believes its device will pay for itself in less than a year, and it plans to target not just restaurants but also convenience stores, food trucks, and perhaps even vending machines. The company argues that eliminating labor costs and reducing the amount of space required in kitchens will allow restaurants to spend more on high-quality ingredients, enabling them to offer gourmet hamburgers at fast food prices.

Those burgers might sound very inviting, but they would come at a considerable cost. Millions of people hold low-wage, often part-time, jobs in the fast food and beverage industries. McDonald's alone employs about 1.8 million workers in 34,000 restaurants worldwide.[15] Historically, low wages, few benefits, and a high turnover rate have helped to make fast food jobs relatively easy to find, and fast food jobs, together with other low-skill positions in retail, have provided a kind of private sector safety net for workers with few other options: these jobs have traditionally offered an income of last resort when no better alternatives are available. In December 2013, the US Bureau of Labor Statistics ranked "combined food preparation and serving workers," a category that excludes waiters and waitresses in full-service restaurants, as one of the top employment sectors in terms of the number of job openings projected over the course of the decade leading up to 2022—with nearly half a million new jobs and another million openings to replace workers who leave the industry.[16]

In the wake of the Great Recession, however, the rules that used to apply to fast food employment are changing rapidly. In 2011, McDonald's launched a high-profile initiative to hire 50,000 new workers in a single day and received over a million applications—a ratio that made landing a McJob more of a statistical long shot than getting accepted at Harvard. While fast food employment was once dominated by young people looking for a part-time income while in school, the industry now employs far more mature workers who rely on the jobs as their primary income. Nearly 90 percent of fast food workers are twenty or older, and the average age is thirty-five.[17] Many of these older workers have to support families—a nearly impossible task at a median wage of just $8.69 per hour.

The industry's low wages and nearly complete lack of benefits have drawn intensive criticism. In October 2013, McDonald's was lambasted after an employee who called the company's financial help line was advised to apply for food stamps and Medicaid.[18] Indeed, an analysis by the Labor Center at the University of California, Berkeley, found that more than half of the families of fast food workers are enrolled in some type of public assistance program and that the resulting cost to US taxpayers is nearly $7 billion per year.[19]

When a spate of protests and ad hoc strikes at fast food restaurants broke out in New York and then spread to more than fifty US cities in the fall of 2013, the Employment Policies Institute, a conservative think tank with close ties to the restaurant and hotel industries, placed a full-page ad in the *Wall Street Journal* warning that "Robots Could Soon Replace Fast Food Workers Demanding a Higher Minimum Wage." While the ad was doubtless intended as a scare tactic, the reality is that—as the Momentum Machines device demonstrates—increased automation in the fast food industry is almost certainly inevitable. Given that companies like Foxconn are introducing robots to perform high-precision electronic assembly in China, there is little reason to believe that machines won't also eventually be serving up burgers, tacos, and lattes across the fast food industry.*

Japan's Kura sushi restaurant chain has already successfully pioneered an automation strategy. In the chain's 262 restaurants, robots

---

* Economists categorize fast food as part of the service sector; however, from a technical standpoint it is really closer to being a form of just-in-time manufacturing.

help make the sushi while conveyor belts replace waiters. To ensure freshness, the system keeps track of how long individual sushi plates have been circulating and automatically removes those that reach their expiration time. Customers order using touch panel screens, and when they are finished dining they place the empty dishes in a slot near their table. The system automatically tabulates the bill and then cleans the plates and whisks them back to the kitchen. Rather than employing store managers at each location, Kura uses centralized facilities where managers are able to remotely monitor nearly every aspect of restaurant operations. Kura's automation-based business model allows it to price sushi plates at just 100 yen (about $1), significantly undercutting its competitors.[20]

It's fairly easy to envision many of the strategies that have worked for Kura, especially automated food production and offsite management, eventually being adopted across the fast food industry. Some significant steps have already been taken in that direction; McDonalds, for example, announced in 2011 that it would install touch screen ordering systems at 7,000 of its European restaurants.[21] Once one of the industry's major players begins to gain significant advantages from increased automation, the others will have little choice but to follow suit. Automation will also offer the ability to compete on dimensions beyond lower labor costs. Robotic production might be viewed as more hygienic since fewer workers would come into contact with the food. Convenience, speed, and order accuracy would increase, as would the ability to customize orders. Once a customer's preferences were recorded at one restaurant, automation would make it a simple matter to consistently produce the same results at other locations.

Given all this, I think it is quite easy to imagine that a typical fast food restaurant may eventually be able to cut its workforce by 50 percent, or perhaps even more. At least in the United States, the fast food market is already so saturated that it seems very unlikely that new restaurants could make up for such a dramatic reduction in the number of workers required at each location. And this, of course, would mean that a great many of the job openings forecast by the Bureau of Labor Statistics might never materialize.

The other major concentration of low-wage service jobs is in the general retail sector. Economists at the Bureau of Labor Statistics rank "retail salesperson" second only to "registered nurse" as the specific occupation that will add the most jobs in the decade ending in 2020 and expect over 700,000 new jobs to be created.[22] Once again, however, technology has the potential to make the government projections seem optimistic. We can probably anticipate that three major forces will shape employment in the retail sector going forward.

The first will be the continuing disruption of the industry by online retailers like Amazon, eBay, and Netflix. The competitive advantage that online suppliers have over brick and mortar stores is already, of course, evident with the demise of major retail chains like Circuit City, Borders, and Blockbuster. Both Amazon and eBay are experimenting with same-day delivery in a number of US cities, with the objective of undermining one of the last major advantages that local retail stores still enjoy: the ability to provide immediate gratification after a purchase.

In theory, the encroachment of online retailers should not necessarily destroy jobs but, rather, would transition them from traditional retail settings to the warehouses and distribution centers used by the online companies. However, the reality is that once jobs move to a warehouse they become far easier to automate. Amazon purchased Kiva Systems, a warehouse robotics company in 2012. Kiva's robots, which look a bit like huge, roving hockey pucks, are designed to move materials within warehouses. Rather than having workers roam the aisles selecting items, a Kiva robot simply zips under an entire pallet or shelving unit, lifts it, and then brings it directly to the worker packing an order. The robots navigate autonomously using a grid laid out by barcodes attached to the floor and are used to automate warehouse operations at a variety of major retailers in addition to

Amazon, including Toys "R" Us, the Gap, Walgreens, and Staples.[23] A year after the acquisition, Amazon had about 1,400 Kiva robots in operation but had only begun the process of integrating the machines into its massive warehouses. One Wall Street analyst estimates that the robots will ultimately allow the company to cut its order fulfillment costs by as much as 40 percent.[24]

The Kroger Company, one of the largest grocery retailers in the United States, has also introduced highly automated distribution centers. Kroger's system is capable of receiving pallets containing large supplies of a single product from vendors and then disassembling them and creating new pallets containing a variety of different products that are ready to ship to stores. It is also able to organize the way that products are stacked on the mixed pallets in order to optimize the stocking of shelves once they arrive at stores. The automated warehouses completely eliminate the need for human intervention, except for loading and unloading the pallets onto trucks.[25] The obvious impact that these automated systems have on jobs has not been lost on organized labor, and the Teamsters Union has repeatedly clashed with Kroger, as well as other grocery retailers, over their introduction. Both the Kiva robots and Kroger's automated system do leave some jobs for people, and these are primarily in areas, such as packing a mixture of items for final shipment to customers, that require visual recognition and dexterity. Of course, these are the very areas in which innovations like Industrial Perception's box-moving robots are rapidly advancing the technical frontier.

The second transformative force is likely to be the explosive growth of the fully automated self-service retail sector—or, in other words, intelligent vending machines and kiosks. One study projects that the value of products and services vended in this market will grow from about $740 billion in 2010 to more than $1.1 trillion by 2015.[26] Vending machines have progressed far beyond dispensing sodas, snacks, and lousy instant coffee, and sophisticated machines that sell consumer electronics products like Apple's iPod and iPad are now common in airports and upscale hotels. AVT, Inc., one of the leading manufacturers of automated retail machines, claims that it can design a custom self-service solution for virtually any product. Vending machines make it possible to dramatically reduce three of the most significant costs incurred in the retail business: real estate, labor, and theft by customers and employees. In addition to providing 24-hour service, many of the machines include video screens and are able to offer targeted point-of-sale advertising that's geared toward enticing customers to purchase related products in much the same way that a human sales clerk might do. They can also collect customer email addresses and send receipts. In essence, the machines offer many of the advantages of online ordering, with the added benefit of instant delivery.

While the proliferation of vending machines and kiosks is certain to eliminate traditional retail sales jobs, these machines will also, of course, create jobs in areas like maintenance, restocking, and repair. The number of those new jobs, however, is likely to be more limited than you might expect. The latest-generation machines are directly connected to the Internet and provide a continuous stream of sales and diagnostic data; they are also specifically designed to minimize the labor costs associated with their operation.

In 2010, David Dunning was the regional operations supervisor responsible for overseeing the maintenance and restocking of 189 Redbox movie rental kiosks in the Chicago area.[27] Redbox has over 42,000 kiosks in the United States and Canada, typically located at convenience stores and supermarkets, and rents about 2 million videos per day.[28] Dunning managed the Chicago-area kiosks with a staff of just seven. Restocking the machines is highly automated; in fact, the most labor-intensive aspect of the job is swapping the translucent movie advertisements displayed on the kiosk—a process that typically takes less than two minutes for each machine. Dunning and his staff divide their time between the warehouse, where new movies arrive, and their cars and homes, where they are able to access and manage the machines via the Internet. The kiosks are designed from the ground up for remote maintenance. For example, if a machine jams it will report this immediately, and a technician can log in with his or her laptop computer, jiggle the mechanism, and fix the problem without the need to visit the site. New movies are typically released on Tuesdays, but the machines can be restocked at any time prior to that; the kiosk will automatically make the movies available for rental at the right time. That allows technicians to schedule restocking visits to avoid traffic.

While the jobs that Dunning and his staff have are certainly interesting and desirable, in number they are a fraction of what a traditional retail chain would create. The now-defunct Blockbuster, for example, once had dozens of stores in greater Chicago, each employing its own sales staff.[29] At its peak, Blockbuster had a total of about 9,000 stores and 60,000 employees. That works out to about seven jobs per store—roughly the same number that Redbox employed in the entire region serviced by Dunning's team.

The third major force likely to disrupt employment in the retail sector will be the introduction of increased automation and robotics into stores as brick and mortar retailers strive to remain competitive. The same innovations that are enabling manufacturing robots to advance the frontier in areas like physical dexterity and visual recognition will eventually allow retail automation to begin moving from warehouses into more challenging and varied environments like stocking shelves in stores. In fact, as far back as 2005, Walmart was already investigating the possibility of using robots that rove store aisles at night and automatically scan barcodes in order to track product inventories.[30]

At the same time, self-service checkout aisles and in-store information kiosks are sure to become easier to use, as well as more common. Mobile devices will also become an ever more important self-service tool. Future shoppers will rely more and more on their phones as a way to shop, pay, and get help and information about products while in traditional retail settings. The mobile disruption of retail is already under way. Walmart, for example, is testing an experimental program that allows shoppers to scan barcodes and then checkout and pay with their phones—completely avoiding long checkout lines.[31] Silvercar, a start-up rental car company, offers the capability to reserve and pick up a car without ever having to interact with a rental clerk; the customer simply scans a barcode to unlock the car and then drives away.[32] As natural language technology like Apple's Siri or even more powerful systems like IBM's Watson continue to advance and become more affordable, it's easy to imagine shoppers soon being able to ask their mobile devices for assistance in much the same way they might ask a store employee. The difference, of course, is that the customer will never have to wait for or hunt down the employee; the virtual assistant will always be instantly available and will rarely, if ever, give an inaccurate answer.

While many retailers may choose to bring automation into traditional retail configurations, others may instead elect to entirely redesign stores—perhaps, in essence, turning them into scaled-up vending machines. Stores of this type might consist of an automated warehouse with an attached showroom where customers could examine product samples and place orders. Orders might then be delivered directly to customers, or perhaps even loaded robotically into vehicles. Regardless of the specific technological path ultimately followed by the retail industry, it's difficult to imagine that the eventual result won't be more robots and machines—and significantly fewer jobs for people.

Cloud Robotics

One of the most important propellants of the robot revolution may turn out to be "cloud robotics"—or the migration of much of the intelligence that animates mobile robots into powerful, centralized computing hubs. Cloud robotics has been enabled by the dramatic acceleration in the rate at which data can be communicated; it is now possible to offload much of the computation required by advanced robotics into huge data centers while also giving individual robots access to network-wide resources. That, of course, makes it possible to build less expensive robots, since less onboard computational power and memory are required, and also allows for instant software upgrades across multiple machines. If one robot employs centralized machine intelligence to learn and adapt to its environment, then that newly acquired knowledge could become instantly available to any other machines accessing the system—making it easy to scale machine learning across large numbers of robots. Google announced support for cloud robotics in 2011 and provides an interface that allows robots to take advantage of all the services designed for Android devices.*

The impact of cloud robotics may be most dramatic in areas like visual recognition that require access to vast databases as well as powerful computational capability. Consider, for example, the enormous technical challenge involved in building a robot capable of performing a variety of housekeeping chores. A robotic maid tasked with clearing up the clutter in a room would need to be able to recognize an almost unlimited number of objects and then decide what to do with them. Each of those items might come in a variety of styles, be oriented in different ways, and perhaps even be somehow entangled with other objects. Compare that challenge to the one taken on by the Industrial Perception box-moving robot we met at the beginning of this chapter. While that robot's ability to discern and grasp individual boxes even when they are stacked in a careless way is an impressive achievement, it is still limited to, well, boxes. That's obviously a very long way from being able to recognize and manipulate virtually any object of any shape and in any configuration.

* Google's strong interest in robotics was further demonstrated in 2013, when the company purchased eight robotics start-up companies over a six-month period. Among the companies acquired was Industrial Perception.

Building such comprehensive visual perception and recognition into an affordable robot poses a daunting challenge. Yet, cloud robotics offers at least a glimpse of the path that may eventually lead to a solution. Google introduced its "Goggles" feature for camera-equipped mobile devices in 2010 and has significantly improved the technology since then. This feature allows you to take a photo of things like landmark buildings, books, works of art, and commercial products and then have the system automatically recognize and retrieve information relevant to the photo. While building the ability to recognize nearly any object into a robot's onboard system would be extraordinarily difficult and expensive, it's fairly easy to imagine robots of the future recognizing the objects in their environment by accessing a vast centralized database of images similar to the one used by the Goggles system. The cloud-based image library could be updated continuously, and any robots with access to the system would get an instant upgrade to their visual recognition capability.

Cloud robotics is sure to be a significant driver of progress in building more capable robots, but it also raises important concerns, especially in the area of security. Aside from its uncomfortable similarity to "Skynet," the controlling machine intelligence in the *Terminator* movies starring Arnold Schwarzenegger, there is the much more practical and immediate issue of susceptibility to hacking or cyber attack. This will be an especially significant concern if cloud robotics someday takes on an important role in our transportation infrastructure. For example, if automated trucks and trains eventually move food and other critical supplies under centralized control, such a system might create extreme vulnerabilities. There is already great concern about the vulnerability of industrial machinery, and of vital infrastructure like the electrical grid, to cyber attack. That vulnerability was demonstrated by the Stuxnet worm that was created by the US and Israeli governments in 2010 to attack the centrifuges used in Iran's nuclear program. If, someday, important infrastructure components are dependent on centralized machine intelligence, those concerns could be raised to an entirely new level.

Robots in Agriculture

Of all the employment sectors that make up the US economy, agriculture stands out as the one that has already undergone the most dramatic transformation as a direct result of technological progress. Most of those new technologies were, of course, mechanical in nature and came long before the advent of advanced information technology. In the late nineteenth century, nearly half of all US workers were employed on farms; by 2000 that fraction had fallen below 2 percent. For crops like wheat, corn, and cotton that can be planted, maintained, and harvested mechanically, the human labor required per bushel of output is now nearly negligible in advanced countries. Many aspects of raising and managing livestock are also mechanized. For example, robotic milking systems are in common use on dairy farms, and in the United States, chickens are grown to standardized sizes so as to make them compatible with automated slaughtering and processing.

The remaining labor-intensive areas of agriculture are primarily geared toward picking delicate, high-value fruits and vegetables, as well as ornamental plants and flowers. As with other relatively routine, manual occupations, these jobs have so far been protected from mechanization primarily because they are highly dependent on visual perception and dexterity. Fruits and vegetables are easily damaged and often need to be selected based on color or softness. For a machine, visual recognition is a significant challenge: lighting conditions can be highly variable, and individual fruits can be in a variety of orientations and may be partly or even completely obscured by leaves.

The same innovations that are advancing the robotics frontier in factory and warehouse settings are finally making many of these remaining agricultural jobs susceptible to automation. Vision

Robotics, a company based in San Diego, California, is developing an octopus-like orange harvesting machine. The robot will use three-dimensional machine vision to make a computer model of an entire orange tree and then store the location of each fruit. That information will then be passed on to the machine's eight robotic arms, which will rapidly harvest the oranges.[33] Boston-area start-up Harvest Automation is initially focused on building robots to automate operations in nurseries and greenhouses; the company estimates that manual labor accounts for over 30 percent of the cost of growing ornamental plants. In the longer run, the company believes that its robots will be able to perform up to 40 percent of the manual agricultural labor now required in the United States and Europe.[34] Experimental robots are already pruning grapevines in France using machine vision technology combined with algorithms that decide which stems should be cut.[35] In Japan, a new machine is able to select ripe strawberries based on subtle color variations and then pick a strawberry every eight seconds—working continuously and doing most of the work at night.[36]

Advanced agricultural robots are especially attractive in countries that do not have access to low-wage, migrant labor. Australia and Japan, for example, are both island nations with rapidly aging workforces. Security considerations likewise make Israel a virtual island in terms of labor mobility. Many fruits and vegetables need to be harvested within a very small time window, so that a lack of available workers at just the right time can easily turn out to be a catastrophic problem.

Beyond reducing the need for labor, agricultural automation has enormous potential to make farming more efficient and far less resource-intensive. Computers have the ability to track and manage crops at a level of granularity that would be inconceivable for human workers. The Australian Centre for Field Robotics (ACFR) at the University of Sydney is focused on employing advanced agricultural robotics to help position Australia as a primary supplier of food for Asia's exploding population—in spite of the country's relative paucity of arable land and fresh water. ACFR envisions robots that continuously prowl fields taking soil samples around individual plants and then injecting just the right amount of water or fertilizer.[37] Precision application of fertilizer or pesticides to individual plants, or even to specific fruits growing on a tree, could potentially reduce the use of these chemicals by up to 80 percent, thereby dramatically decreasing the amount of toxic runoff that ultimately ends up fouling rivers, streams, and other bodies of water.[38] *

Agriculture in most developing countries is notoriously inefficient. The plots of land worked by families are often tiny, capital investment is minimal, and modern technology is unavailable. Even though farming techniques are labor-intensive, the land often has to support more people than are really necessary to cultivate it. As global population grows to 9 billion and beyond in the coming decades, there will be ever-increasing pressure to transition any and all available arable land into larger and more efficient farms that are capable of producing higher crop yields. Advancing agricultural technology will have a significant role to play, especially in countries where water is scarce and ecosystems have been damaged by overuse of chemicals. Increased mechanization, however, will also mean that the land will provide livelihoods for far fewer people. The historical norm has been for those excess workers to migrate to cities and industrial centers in search of factory work—but as we have seen, those factories are themselves going to be transformed by accelerating automation technology. In fact, it seems somewhat difficult to imagine how many developing countries will succeed in navigating these technological disruptions without running into significant unemployment crises.

---

* Precision agriculture—or the ability to keep track of and manage individual plants or even fruits—is part of the "big data" phenomenon, a subject that we'll examine in more depth in Chapter 4.

In the United States, agricultural robotics has the potential to eventually throw a wrench into many of the fundamental assumptions that underlie immigration policy—an area that is already subject to intensely polarized politics. The impact is already evident in some areas that used to employ large numbers of farmworkers. In California, machines skirt around the daunting visual challenge of picking individual almonds by simply grasping the entire tree and violently shaking it. The almonds fall to the ground where they'll be harvested by a different machine. Many California farmers have transitioned from delicate crops like tomatoes to more robust nuts because they can be harvested mechanically. Overall agricultural employment in California fell by about 11 percent in the first decade of the twenty-first century, even as the total production of crops like almonds, which are compatible with automated farming techniques, has exploded.[39]

As robotics and advanced self-service technologies are increasingly deployed across nearly every sector of the economy, they will primarily threaten lower-wage jobs that require modest levels of education and training. These jobs, however, currently make up the vast majority of the new positions being generated by the economy—and the US economy needs to create something on the order of a million jobs per year just to tread water in the face of population growth. Even if we set aside the possibility of an actual reduction in the number of these jobs as new technologies emerge, any decline in the rate at which they are created will have dire, cumulative consequences for employment over the long run.

Many economists and politicians might be inclined to dismiss this as a problem. After all, routine, low-wage, low-skill jobs—at least in advanced economies—tend to be viewed as inherently undesirable, and when economists discuss the impact of technology on these kinds of jobs, you are very likely to encounter the phrase "freed up"—as in, workers who lose their low-skill jobs will be freed up to pursue more training and better opportunities. The fundamental assumption, of course, is that a dynamic economy like the United States will always be capable of generating sufficient higher-wage, higher-skill jobs to absorb all those newly freed up workers—given that they succeed in acquiring the necessary training.

That assumption rests on increasingly shaky ground. In the next two chapters we'll look at the impact that automation has already had on jobs and incomes in the United States and consider the characteristics that set information technology apart as a uniquely disruptive force. That discussion will provide a jumping-off point from which to delve into an unfolding story that is poised to upend the conventional wisdom about the types of jobs most likely to be automated and the viability of ever more education and training as a solution: the machines are coming for the high-wage, high-skill jobs as well.

Chapter 2

IS THIS TIME DIFFERENT?

On the morning of Sunday, March 31, 1968, the Reverend Martin Luther King, Jr., stood in the elaborately carved limestone pulpit at Washington National Cathedral. The building—one of the largest churches in the world and over twice the size of London's Westminster abbey—was filled to capacity with thousands of people packed into the nave and transept, looking down from the choir loft, and squeezed into doorways. At least another thousand people gathered outside on the steps or at nearby St. Alban's Episcopal Church to hear the sermon over loudspeakers.

It would be Dr. King's final Sunday sermon. Just five days later the cathedral would again be overflowing with a far more somber crowd—including President Lyndon Johnson, senior cabinet officials, all nine Supreme Court justices, and leading members of Congress—gathered to honor King at a memorial service the day following his assassination in Memphis, Tennessee.[1]

The title of Dr. King's sermon that day was "Remaining Awake Through a Great Revolution." Civil and human rights were, as might be expected, a major component of his address, but he had in mind revolutionary change on a much broader front. As he explained a short way into his sermon:

> There can be no gainsaying of the fact that a great revolution is taking place in the world today. In a sense it is a triple revolution: that is, a technological revolution, with the impact of automation and cybernation; then there is a revolution in weaponry, with the emergence of atomic and nuclear weapons of warfare; then there is a human rights revolution, with the freedom explosion that is taking place all over the world. Yes, we do live in a period where changes are taking place. And there is still the voice crying through the vista of time saying, "Behold, I make all things new; former things are passed away."[2]

The phrase "triple revolution" referred to a report written by a group of prominent academics, journalists, and technologists that called itself the Ad Hoc Committee on the Triple Revolution. The group included Nobel laureate chemist Linus Pauling as well as economist Gunnar Myrdal, who would be awarded the Nobel Prize in economics, along with Friedrich Hayek, in 1974. Two of the revolutionary forces identified in the report—nuclear weapons and the civil rights movement—are indelibly woven into the historical narrative of the 1960s. The third revolution, which comprised the bulk of the document's text, has largely been forgotten. The report predicted that "cybernation" (or automation) would soon result in an economy where "potentially unlimited output can be achieved by systems of machines which will require little cooperation from human beings."[3] The result would be massive unemployment, soaring inequality, and, ultimately, falling demand for goods and services as consumers increasingly lacked the purchasing power necessary to continue driving economic growth. The Ad Hoc Committee went on to propose a radical solution: the eventual implementation of a guaranteed minimum income made possible by the "economy of abundance" such widespread automation could create, and which would "take the place of the patchwork of welfare measures" that were then in place to address poverty.*

The Triple Revolution report was released to the media and sent to President Johnson, the secretary of labor, and congressional leaders in March 1964. An accompanying cover letter warned ominously that if something akin to the report's proposed solutions was not implemented, "the nation will be thrown into unprecedented economic and social disorder." A front-page story with extensive quotations from the report appeared in the next day's *New York Times,* and numerous other newspapers and magazines ran stories and editorials (most of which were critical), in some cases even printing the entire text of the report.[4]

The Triple Revolution marked what was perhaps the crest of a wave of worry about the impact of automation that had arisen following World War II. The specter of mass joblessness as machines displaced workers had incited fear many times in the past—going all the way back to Britain's Luddite uprising in 1812—but in the 1950s and '60s, the concern was especially acute and was articulated by some of the United States' most prominent and intellectually capable individuals.

In 1949, at the request of the *New York Times,* Norbert Wiener, an internationally renowned mathematician at the Massachusetts Institute of Technology, wrote an article describing his vision for the future of computers and automation.[5] Wiener had been a child prodigy who entered college at age eleven and completed his PhD

---

* The Committee on the Triple Revolution did not advocate the immediate implementation of a guaranteed income. Instead, it proposed a list of nine transitional policies. Many of these were quite conventional, and included things such as greatly increased investment in education, public works projects to create jobs, and the construction of low-cost housing. The report also argued for a greatly expanded role for unions and suggested that organized labor should become an advocate for the unemployed as well as those who held jobs.

when he was just seventeen; he went on to establish the field of cybernetics and made substantial contributions in applied mathematics and to the foundations of computer science, robotics, and computer-controlled automation. In his article—written just three years after the first true general-purpose electronic computer was built at the University of Pennsylvania*—Wiener argued that "if we can do anything in a clear and intelligible way, we can do it by machine" and warned that that this could ultimately lead to "an industrial revolution of unmitigated cruelty" powered by machines capable of "reducing the economic value of the routine factory employee to a point at which he is not worth hiring at any price."**

Three years later, a dystopian future much like the one Wiener had imagined was brought to life in the pages of Kurt Vonnegut's first novel. *Player Piano* described an automated economy in which industrial machines managed by a tiny technical elite did virtually all the work, while the vast majority of the population faced a meaningless existence and a hopeless future. Vonnegut, who went on to achieve legendary status as an author, continued to believe in the relevance of his 1952 novel throughout his life, writing decades later that it was becoming "more timely with each passing day."[6]

Four months after the Johnson administration received the Triple Revolution report, the president signed a bill creating the National Commission on Technology, Automation, and Economic Progress.[7] In his remarks at the bill's signing ceremony, Johnson said that "automation can be the ally of our prosperity if we will just

---

* ENIAC (Electronic Numerical Integrator and Computer) was built at the University of Pennsylvania in 1946. A true programmable computer, it was financed by the US Army and intended primarily for calculating firing tables used to aim artillery.

** Due to a miscommunication, Wiener's article was never published in 1949. A draft copy was discovered by a researcher working with documents in the MIT library archives in 2012, and substantial excerpts were finally published in a May 2013 article by *New York Times* science reporter John Markoff.

look ahead, if we will understand what is to come, and if we will set our course wisely after proper planning for the future." The newly formed commission then—as is almost universally the case with such commissions—quickly faded into obscurity, leaving behind at least three book-length reports of its own.[8]

The irony of all the automation worries in the postwar period was that the economy offered very little in the way of evidence to support such concerns. When the Triple Revolution report was released in 1964, the unemployment rate was just over 5 percent, and it would fall to a low of 3.5 percent by 1969. Even during the four recessions that occurred between 1948 and 1969, unemployment never reached 7 percent, and then it fell rapidly once recovery was under way.[9] The introduction of new technologies did drive substantial increases in productivity, but the lion's share of that growth was captured by workers in the form of higher wages.

By the early 1970s, focus had shifted to the OPEC oil embargo, and then to the subsequent years of stagflation. The potential for machines and computers to cause unemployment was pushed further and further out of the mainstream. Among professional economists in particular, the idea became virtually untouchable. Those who did dare to entertain such thoughts risked being labeled a "neo-Luddite."

Given that the dire circumstances predicted by the Triple Revolution report did not come to pass, we can ask an obvious question: Were the authors of the report definitively wrong? Or did they—like many others before them—simply sound the alarm far too soon?

Norbert Wiener, as one of the early pioneers of information technology, perceived the digital computer as being fundamentally different from the mechanical technologies that preceded it. It was a game changer: a new kind of machine with the potential to usher in a new age—and, ultimately, perhaps rend the very fabric of society. Yet, Wiener's views were expressed at a time when computers were room-sized monstrosities whose calculations were powered by tens of thousands of searingly hot radio vacuum tubes, some number of which could be expected to fail on a near daily basis.[10] It would be decades before the exponential arc of progress would drive digital technology to a level where such views might reasonably be justified.

Those decades are now behind us, and the time is ripe for an open-minded reassessment of the impact of technology on the economy. The data shows that even as concerns about the impact of labor-saving technology receded to the fringes of economic thought, something that had been fundamental to the postwar era of prosperity gradually began to change in the American economy. The nearly perfect historical correlation between increasing productivity and rising incomes broke down: wages for most Americans stagnated and, for many workers, even declined; income inequality soared to levels not seen since the eve of the 1929 stock market crash; and a new phrase—"jobless recovery"—found a prominent place in our vocabulary. In all, we can enumerate at least seven economic trends that, taken together, suggest a transformative role for advancing information technology.

Seven Deadly Trends

Stagnant Wages

The year 1973 was an eventful one in the history of the United States. The Nixon administration was embroiled in the Watergate scandal, and in October, OPEC initiated an oil embargo that would soon result in long lines of angry motorists at gas stations across the country. Even as Nixon descended into his death spiral, however, there was another story unfolding. This story began with an event that went completely unheralded and yet marked the beginning of a trend that would arguably dwarf both Watergate and the oil crisis in importance. For that was the year a typical American worker's pay reached its peak. Measured in 2013 dollars, a typical worker—that is, production and nonsupervisory workers in the private sector, representing well over half the American workforce—earned about $767 per week in

1973. The following year, real average wages began a precipitous decline from which they would never fully recover. A full four decades later, a similar worker earns just $664, a decline of about 13 percent.[11]

The story is modestly better if we look at median household incomes. Between 1949 and 1973, US median household incomes roughly doubled, from about $25,000 to $50,000. Growth in median incomes during this period tracked nearly perfectly with per capita GDP. Three decades later, median household income had increased to about $61,000, an increase of just 22 percent. That growth, however, was driven largely by the entry of women into the workforce. If incomes had moved in lockstep with economic growth—as was the case prior to 1973—the median household would today be earning well in excess of $90,000, over 50 percent more than the $61,000 they do earn.[12]

Figure 2.1 shows the relationship between labor productivity* (which measures the value of workers' hourly output) and compensation (which includes wages and benefits) paid to ordinary private sector workers from 1948 onward. The first segment of the graph (from 1948 to 1973) shows the way economists expect things to work. Growth in productivity moves in almost perfect lockstep with compensation. Prosperity marches upward and is shared broadly by all those who contribute to the economy. Beyond the mid-1970s, the widening gap between the two lines is a graphic illustration of the extent to which the fruits of innovation throughout the economy are now accruing almost entirely to business owners and investors, rather than to workers.

---

* Labor productivity measures the value of the output (either goods or services) produced by workers per hour. It is a critically important gauge of the general efficiency of an economy; to a significant extent it determines the wealth of a nation. Advanced, industrialized countries have high productivity because their workers have access to more and better technology, enjoy better nutrition as well as safer and more healthful environments, and are generally better educated and trained. Poor countries lack these things and are, therefore, less productive; their people must work longer and harder to produce the same level of output.

Figure 2.1. Growth of Real Hourly Compensation for Production and Nonsupervisory Workers Versus Productivity (1948–2011)

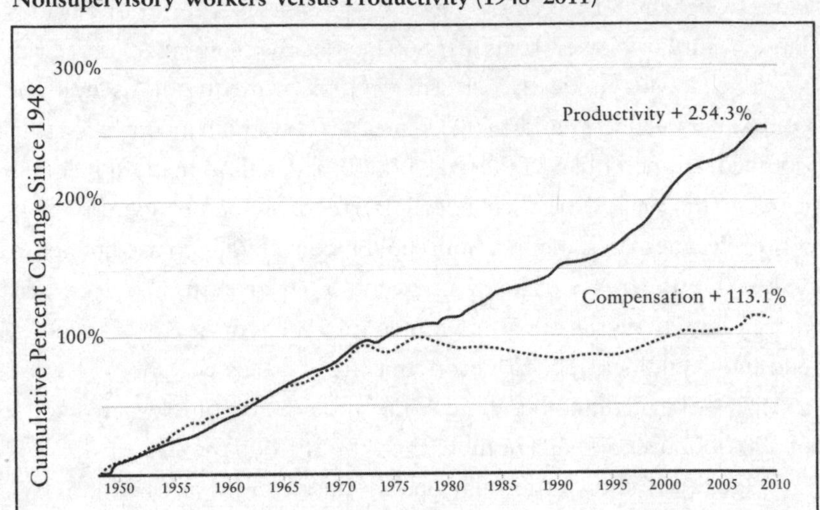

SOURCE: Lawrence Mishel, Economic Policy Institute, based on an analysis of unpublished total economy data from the Bureau of Labor Statistics, the Labor Productivity and Costs program, and the Bureau of Economic Analysis's National Income and Product Accounts public data series.[13]

Despite the clarity of this graph, many economists have still not fully acknowledged the divergence between wage and productivity growth. Figure 2.2 shows how growth rates for compensation and productivity compare during different periods going back to 1947. Productivity has significantly outstripped compensation in every decade from 1980 on. The difference is especially dramatic from 2000 to 2009; although productivity growth nearly matches the 1947–1973 period—the golden era of postwar prosperity—compensation lags far behind. It's difficult to look at this graph and not come away with the impression that productivity growth is pretty clearly blowing the doors off the raises that most workers are getting.

The authors of most college economics textbooks have been especially slow to acknowledge this picture. Consider, for example,

Figure 2.2. Productivity Growth Versus Compensation Growth

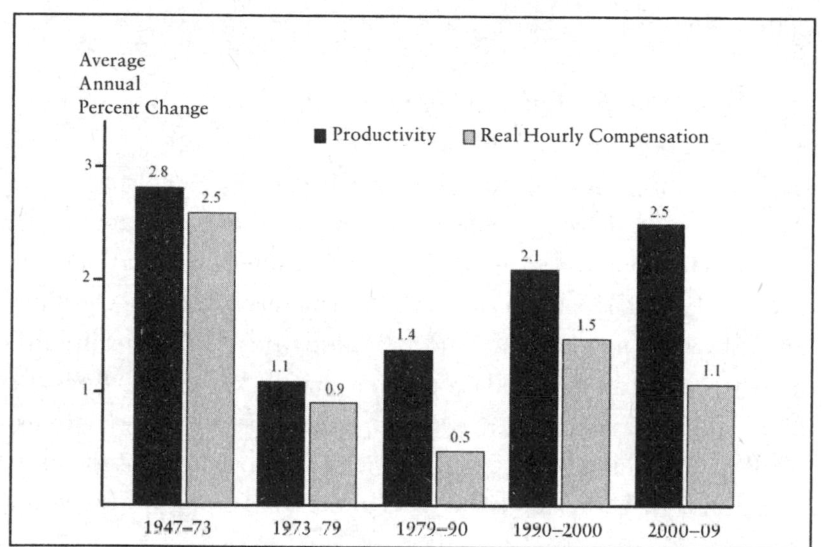

SOURCE: US Bureau of Labor Statistics.[14]

Principles of Economics, an introductory textbook authored by John B. Taylor and Akila Weerapana,[15] the required text for Professor Taylor's wildly popular introductory economics class at Stanford University. It includes a bar chart very similar to Figure 2.2, but still argues for a tight relationship between wages and productivity. What about the fact that productivity leaps away from wages beginning in the 1980s? Taylor and Weerapana note that "the relationship is not perfect." That appears to be something of an understatement. The 2007 edition of another textbook, also titled *Principles of Economics*[16] and co-authored by Princeton professor—and former Federal Reserve chairman—Ben Bernanke, suggests that slow wage growth from 2000 on may have resulted from "the weak labor market that followed the recession of 2001" and that wages ought to "catch up to productivity growth as the labor market returns to normal"—a view that seems to ignore the fact that the tight correlation between wage and productivity growth began to deteriorate long before today's college students were born.*

A Bear Market for Labor's Share, and a Raging Bull for Corporations

Early in the twentieth century, the British economist and statistician Arthur Bowley delved into decades of national income data for the United Kingdom and showed that the fraction of national income going to labor and capital respectively remained relatively constant, at least over long periods. This apparently fixed relationship ultimately became an accepted economic principle known as "Bowley's Law." John Maynard Keynes, perhaps the most famous economist of all time, would later say that Bowley's Law was "one of the most surprising, yet best established facts in the whole range of economic statistics."[17]

As Figure 2.3 shows, during the postwar period, the share of US national income going to labor moved in a fairly tight range, just as

---

* There is also a technical issue that comes into play when discussing the gap between wage growth and productivity growth. Both the wage (or, more broadly, compensation) and productivity numbers must be adjusted for inflation. The standard way to do this, and the method used by the US Bureau of Labor Statistics (BLS), is to use two different measures of inflation. Wages are adjusted using the Consumer Price Index (CPI) because this reflects the prices of products and services that workers actually spend their money on. The productivity figures are adjusted using the GDP deflator (or implicit price deflator), which is a broader measure of inflation in the entire economy. In other words, the GDP deflator incorporates prices for a lot of things that consumers don't actually purchase. One especially important difference is that computers and information technology—which have seen substantial price deflation due to Moore's Law—are much more important in the GDP deflator than in the CPI (computers are not a big component of most household budgets, but are purchased in volume by businesses). Some economists—particularly those who are more conservative—argue that the GDP deflator should be used for *both* wages and productivity. When this method is used, the gap between wage growth and productivity growth narrows significantly. However, this approach almost certainly understates the level of inflation that impacts wage earners.

Figure 2.3. US Labor's Share of National Income (1947–2014)

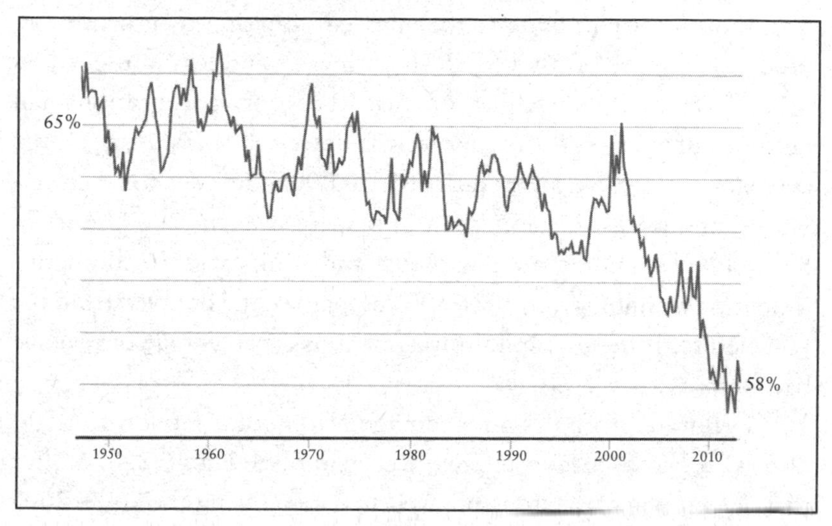

SOURCE: US Bureau of Labor Statistics and Federal Reserve Bank of St. Louis (FRED).[18]

Bowley's Law would have predicted. From the mid-1970s on, however, Bowley's Law began to fall apart as labor's share went first into a gradual decline and then into a seeming free fall just after the turn of the century. The decline is all the more remarkable when we consider that labor's share includes anyone who draws a paycheck. In other words, the enormous salaries of CEOs, Wall Street executives, superstar athletes, and movie stars are all considered labor, and those, of course, haven't been declining at all: they've been skyrocketing. A graph showing the share of national income accruing to ordinary workers—or, more broadly, the bottom 99 percent of the income distribution—would certainly show an even more precipitous plunge.

While labor's share of income plummeted, the story was very different for corporate profits. In April 2012, the *Wall Street Journal* ran a story entitled "For Big Companies, Life Is Good" that documented the astonishing speed at which corporations recovered from the most severe economic crisis since the Great Depression. While millions of workers remained unemployed or accepted jobs at lower pay or with fewer hours, the corporate sector emerged from the downturn "more productive, more profitable, flush with cash and less burdened by debt."[19] Over the course of the Great Recession, corporations had become adept at producing more with fewer workers. In 2011, big companies generated an average of $420,000 in revenue for each employee, an increase of more than 11 percent over the 2007 figure of $378,000.[20] Spending on new plants and equipment, including information technology, by S&P 500 companies had doubled from the year before, bringing capital investment as a percentage of revenue back to pre-crisis levels.

Corporate profits as a percentage of the total economy (GDP) also skyrocketed after the Great Recession (see Figure 2.4). Notice that despite the precipitous plunge in profits during the 2008–2009 economic crisis, the speed at which profitability recovered was unprecedented compared with previous recessions.

Figure 2.4. Corporate Profits as a Percentage of GDP

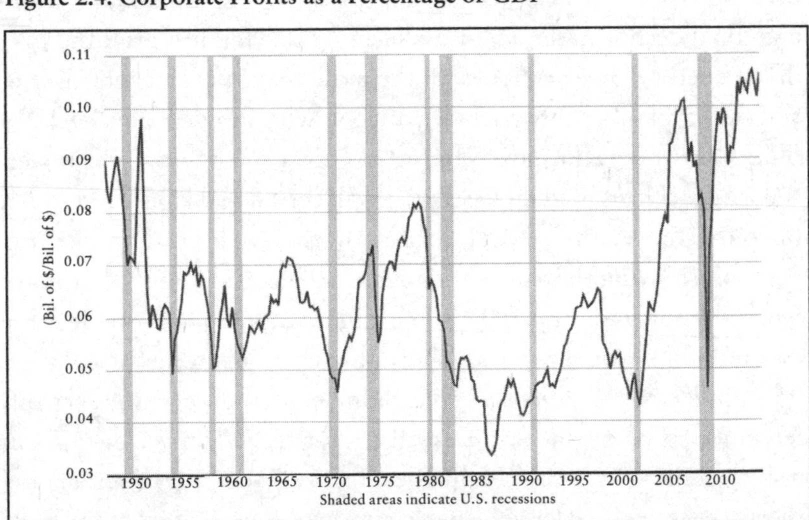

Source: Federal Reserve Bank of St. Louis (FRED).[21]

The decline in labor's share of national income is by no means limited to the United States. In a June 2013 research paper,[22] economists Loukas Karabarbounis and Brent Neiman, both of the University of Chicago's Booth School of Business, analyzed data from fifty-six different countries and found that thirty-eight demonstrated a significant decline in labor's share. In fact, the authors' research showed that Japan, Canada, France, Italy, Germany, and China all had larger declines than the United States over a ten-year period. The decline in labor's share in China—the country that most of us assume is hoovering up all the work—was especially precipitous, falling at three times the rate in the United States.

Karabarbounis and Neiman concluded that these global declines in labor's share resulted from "efficiency gains in capital producing sectors, often attributed to advances in information technology and the computer age."[23] The authors also noted that a stable labor share of income continues to be "a fundamental feature of macroeconomic models."[24] In other words, just as economists do not seem to have fully assimilated the implications of the circa-1973 divergence of productivity and wage growth, they are apparently still quite happy to build Bowley's Law into the equations they use to model the economy.

Declining Labor Force Participation

A separate trend has been the decline in labor force participation. In the wake of the 2008–9 economic crisis, it was often the case that the unemployment rate fell not because large numbers of new jobs were being created, but because discouraged workers exited the workforce. Unlike the unemployment rate, which counts only those people actively seeking jobs, labor-force participation offers a graphic illustration that captures workers who have given up.

As Figure 2.5 shows, the labor force participation rate rose sharply between 1970 and 1990 as women flooded into the workforce. The overall trend disguises the crucial fact that the percentage

Figure 2.5. Labor Force Participation Rate

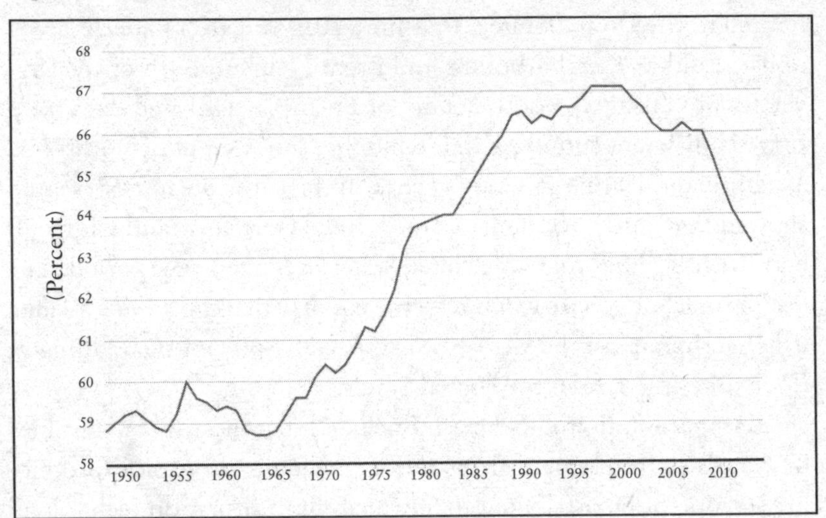

SOURCE: US Bureau of Labor Statistics and Federal Reserve Bank of St. Louis (FRED).[25]

of men in the labor force has been in consistent decline since 1950, falling from a high of about 86 percent to 70 percent as of 2013. The participation rate for women peaked at 60 percent in 2000; the overall labor force participation rate peaked at about 67 percent that same year.[26]

Labor force participation has been falling ever since, and although this is due in part to the retirement of the baby boom generation, and in part because younger workers are pursuing more education, those demographic trends do not fully explain the decline. The labor force participation rate for adults between the ages of twenty-five and fifty-four—those old enough to have completed college and even graduate school, yet too young to retire—has declined from about 84.5 percent in 2000 to just over 81 percent in 2013.[27] In other words, both the overall labor force participation rate and the participation rate for prime working-age adults have fallen by about three percentage points since 2000—and about half of that decline came before the onset of the 2008 financial crisis.

The decline in labor force participation has been accompanied by an explosion in applications for the Social Security disability program, which is intended to provide a safety net for workers who suffer debilitating injuries. Between 2000 and 2011, the number of applications more than doubled, from about 1.2 million per year to nearly 3 million per year.[28] As there is no evidence of an epidemic of workplace injuries beginning around the turn of the century, many analysts suspect that the disability program is being misused as a kind of last-resort—and permanent—unemployment insurance program. Given all this, it seems clear that something beyond simple demographics or cyclical economic factors is driving people out of the labor force.

Diminishing Job Creation, Lengthening Jobless Recoveries, and Soaring Long-Term Unemployment

Over the past half-century, the US economy has become progressively less effective at creating new jobs. Only the 1990s managed to—just barely—keep up with the previous decade's job growth, and that was largely due to the technology boom that occurred in the second half of the decade. The recession that began in December 2007 and the ensuing financial crisis were a total disaster for job creation in the 2000s; the decade ended with virtually the same number of jobs that had existed in December 1999. Even before the Great Recession hit, however, the new century's first decade was already on track to produce by far the worst percentage growth in employment since World War II.

As Figure 2.6 shows, the number of jobs in the economy had increased by only about 5.8 percent through the end of 2007. Prorating that number for the entire decade suggests that, if the economic crisis had not occurred, the 2000s would likely have finished with a roughly 8 percent job creation rate—less than half of the percentage increase seen in the 1980s and '90s.

That miserable job creation performance is especially disturbing in light of the fact that the economy needs to generate large numbers

Figure 2.6. US Job Creation by Decade

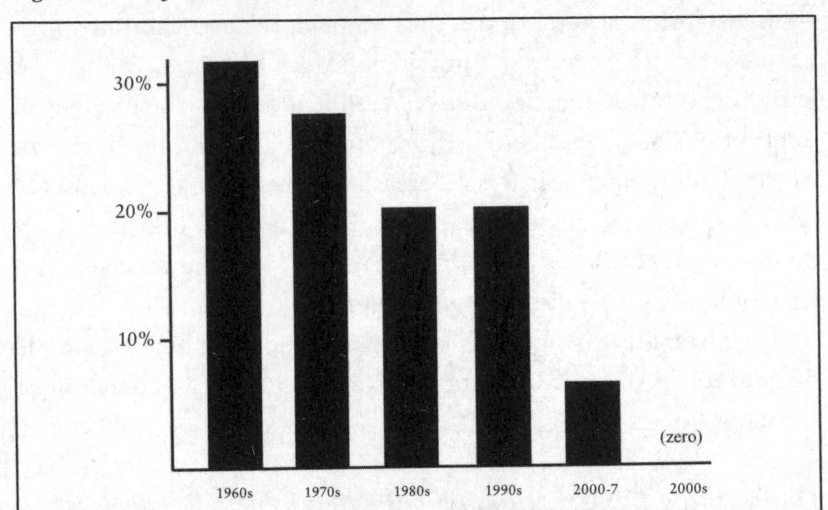

SOURCE: US Bureau of Labor Statistics and Federal Reserve Bank of St. Louis (FRED).[29]

of new jobs—between 75,000 and 150,000 per month, depending on one's assumptions—just to keep up with population growth.[30] Even when the lower estimate is employed, the 2000s still resulted in a deficit of about 9 million jobs over the course of the decade.

Clear evidence also shows that when a recession knocks the wind out of the economy, it is taking longer and longer for the job market to recover. Temporary layoffs have given way to jobless recoveries. A 2010 research report by the Federal Reserve Bank of Cleveland found that recent recessions have seen a dramatic decline in the rate at which unemployed workers are able to land new jobs. In other words, the problem is not that more jobs are being destroyed in downturns; it is that fewer are being created during recoveries. After the onset of the Great Recession in December 2007, the unemployment rate continued to rise for nearly two years, ultimately increasing by a full five percentage points and peaking at 10.1 percent. The Cleveland Fed's analysis found that the increased difficulty faced by workers finding new jobs accounted for over 95 percent of that 5 percent jump in the unemployment rate.[31] This, in turn, has led to a huge jump in the long-term unemployment rate, which peaked in 2010, when about 45 percent of workers had been out of work for more than six months.[32] Figure 2.7 shows the number of months it took for the labor market to recover from recent recessions. The Great Recession resulted in a monstrous jobless recovery; it took until May 2014—a full six and a half years after the start of the downturn—for employment to return to its pre-recession level.

Extended unemployment is a debilitating problem. Job skills erode over time; the risk that workers will become discouraged increases, and many employers seem to actively discriminate against the long-term unemployed, often refusing even to consider their résumés. Indeed, a field experiment conducted by Rand Ghayad, a PhD candidate in economics at Northeastern University, showed

Figure 2.7. US Recessions: Months for Employment to Recover (Measured from Start of Recession)

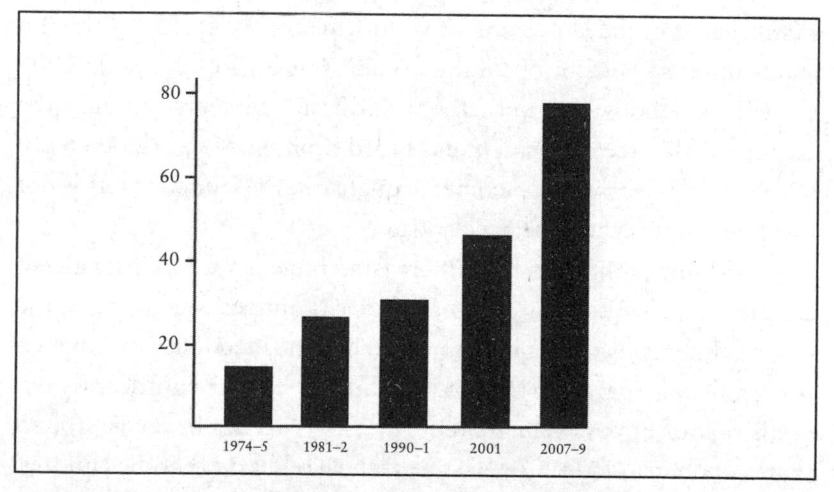

Source: US Bureau of Labor Statistics and Federal Reserve Bank of St. Louis (FRED).[33]

that a recently unemployed applicant with no industry experience was actually more likely to be called in for a job interview than someone with directly applicable experience who had been out of work for more than six months.[34] A separate report by the Urban Institute found that the long-term unemployed are not appreciably different from other workers, suggesting that becoming one of the long-term unemployed—and suffering the stigma that attaches to that category—may largely be a matter of bad luck.[35] If you happen to lose your job at an especially unfavorable time and then fail to find a new position before the dreaded six-month mark (a real possibility if the economy is in free fall), your prospects diminish dramatically from that point on—regardless of how qualified you may be.

Soaring Inequality

The divide between the rich and everyone else has been growing steadily since the 1970s. Between 1993 and 2010 over half of the increase in US national income went to households in the top 1 percent of the income distribution.[36] Since then, things have only gotten worse. In an analysis published in September 2013, economist Emmanuel Saez of the University of California, Berkeley, found that an astonishing 95 percent of total income gains during the years 2009 to 2012 were hoovered up by the wealthiest 1 percent.[37] Even as the Occupy Wall Street movement has faded from the scene, the evidence shows pretty clearly that income inequality in the United States is not just high—it may well be accelerating.

While inequality has been increasing in nearly all industrialized countries, the United States remains a clear outlier. According to the Central Intelligence Agency's analysis, income inequality in America is roughly on a par with that of the Philippines and significantly exceeds that of Egypt, Yemen, and Tunisia.[38] Studies have also found that economic mobility, a measure of the likelihood that the children of the poor will succeed in moving up the income scale, is significantly lower in the United States than in nearly all European nations.

In other words, one of the most fundamental ideas woven into the American ethos—the belief that anyone can get ahead through hard work and perseverance—really has little basis in statistical reality.

From the perspective of any one individual, inequality can be very difficult to perceive. Most people tend to focus their attention locally. They worry about how they are doing relative to the guy next door as opposed to the hedge fund manager they will, in all likelihood, never encounter. Surveys have shown that most Americans vastly underestimate the existing extent of inequality, and when asked to select an "ideal" national distribution of income, they make a choice that, in the real world, exists only in Scandinavian social democracies.[39] *

Nonetheless, inequality has real implications that go far beyond simple frustration about your inability to keep up with the Joneses. Foremost is the fact that the overwhelming success of those at the extreme top seems to be correlated with diminishing prospects for nearly everyone else. The old adage that a rising tide lifts all boats gets pretty tired when you haven't had a meaningful raise since the Nixon administration.

There is also an obvious risk of political capture by the financial elite. In the United States, to a greater degree than in any other advanced democracy, politics is driven almost entirely by money. Wealthy individuals and the organizations they control can mold government policy through political contributions and lobbying, often producing outcomes that are clearly at odds with what the public actually wants. As those at the apex of the income distribution become increasingly detached—living in a kind of bubble that insulates them almost entirely from the realities faced by typical Americans—there

* This is true regardless of political party. In one study conducted by Dan Ariely of Duke University, over 90 percent of Republicans and 93 percent of Democrats preferred an income distribution similar to that of Sweden over that of the United States.

is a real risk that they will be unwilling to support investment in the public goods and infrastructure upon which everyone else depends.

The soaring fortunes of those at the very top may ultimately represent a threat to democratic governance. However, the most immediate problem for most middle- and working-class people is that job market opportunities are broadly deteriorating.

Declining Incomes and Underemployment for Recent College Graduates

A four-year college degree has come to be almost universally viewed as an essential credential for entry into the middle class. As of 2012, average hourly wages for college graduates were more than 80 percent higher than the wages of high school graduates.[40] The college wage premium is a reflection of what economists call "skill biased technological change" (SBTC).* The general idea behind SBTC is that information technology has automated or deskilled much of the work handled by less educated workers, while simultaneously increasing the relative value of the more cognitively complex tasks typically performed by college graduates.

Graduate and professional degrees convey still higher incomes, and in fact, since the turn of the century, things are looking quite a bit less rosy for young college graduates who don't also have an advanced degree. According to one analysis, incomes for young workers with only a bachelor's degree declined nearly 15 percent between 2000 and 2010, and the plunge began well before the onset of the 2008 financial crisis.

---

* SBTC and the college wage premium offer a partial explanation for increasing income inequality. However, since nearly a third of the adult US population has a college degree, if this were the only thing going on, it would imply a much tamer form of inequality than actually exists. The real action is at the very top—and things become more extreme the higher you go. The outsized fortunes of the top 1 (or .01) percent cannot reasonably be attributed to better education or training.

Recent college graduates are also underemployed. By some accounts, fully half of new graduates are unable to find jobs that utilize their education and offer access to the crucial initial rung on the career ladder. Many of these unlucky graduates will probably find it very difficult to move up into solid middle-class trajectories.

To be sure, college graduates have, on average, maintained their income premium over workers with only a high school education, but this is largely because the prospects for these less educated workers have become genuinely dismal. As of July 2013, fewer than half of American workers who were between the ages of twenty and twenty-four and not enrolled in school had full-time jobs. Among nonstudents aged sixteen to nineteen only about 15 percent were working full-time.[41] The return on investment for a college education may be falling, but it still nearly always beats the alternative.

Polarization and Part-Time Jobs

A further new problem is that the jobs being created during economic recoveries are generally worse than those destroyed by recessions. In a 2012 study, economists Nir Jaimovich and Henry E. Siu analyzed data from recent US recessions and found that the jobs mostly likely to permanently disappear are the good middle-class jobs, while the jobs that tend to get created during recoveries are largely concentrated in low-wage sectors like retail, hospitality, and food preparation and, to a lesser extent, in high-skill professions that require extensive training.[42] This has been especially true over the course of the recovery that began in 2009.[43]

Many of these new low-wage jobs are also part-time. Between the start of the Great Recession in December 2007 and August 2013, about 5 million full-time jobs were vaporized, but the number of part-time jobs actually increased by approximately 3 million.[44] That increase in part-time work has occurred entirely among workers who have had their hours cut or who would like a full-time job but are unable to find one.

The propensity for the economy to wipe out solid middle-skill, middle-class jobs, and then to replace them with a combination of low-wage service jobs and high-skill, professional jobs that are generally unattainable for most of the workforce, has been dubbed "job market polarization." Occupational polarization has resulted in an hourglass-shaped job market where workers who are unable to land one of the desirable jobs at the top end up at the bottom.

This polarization phenomenon has been studied extensively by David Autor, an economist at the Massachusetts Institute of Technology. In a 2010 paper, Autor identifies four specific mid-range occupational categories that have been especially hard-hit as polarization has unfolded: sales, office/administrative, production/craft/repair, and operators/fabricators/laborers. Over the thirty years between 1979 and 2009, the percentage of the US workforce employed in these four areas declined from 57.3 percent to 45.7 percent, and there was a noticeable acceleration in the rate of job destruction between 2007 and 2009.[45] Autor's paper also makes it clear that polarization is not limited to the United States, but has been documented in most advanced, industrial economies; in particular, sixteen countries within the European Union have seen a significant decline in the percentage of the workforce engaged in mid-range occupations over the thirteen years between 1993 and 2006.[46]

Autor concludes that the primary driving forces behind job market polarization are "the automation of routine work and, to a smaller extent, the international integration of labor markets through trade and, more recently, offshoring."[47] In their more recent paper showing the relationship between polarization and jobless recoveries, Jaimovich and Siu point out that fully 92 percent of the job losses in mid-range occupations have occurred within a year of a recession.[48] In other words, polarization is not necessarily something that happens according to a grand plan, nor is it a gradual and continuous evolution. Rather, it is an organic process that is deeply intertwined with the business cycle; routine jobs are eliminated for economic reasons during a recession, but organizations then discover that ever-advancing information technology allows them to operate successfully without rehiring the workers once a recovery gets under way. Chrystia Freeland of Reuters puts it especially aptly, writing that "the middle-class frog isn't being gradually boiled; it is being periodically grilled at a very high heat."[49]

A Technology Narrative

It's fairly easy to piece together a hypothetical narrative that puts advancing technology—and the resulting automation of routine work—front and center as the explanation for these seven deadly economic trends. The golden era from 1947 to 1973 was characterized by significant technological progress and strong productivity growth. This was before the age of information technology; the innovations during this period were primarily in areas like mechanical, chemical, and aerospace engineering. Think, for example, of how airplanes evolved from employing internal combustion engines driving propellers to much more reliable and better-performing jet engines. This period exemplified what is written in all those economics textbooks: innovation and soaring productivity made workers more valuable—and allowed them to command higher wages.

In the 1970s, the economy received a major shock from the oil crisis and entered an unprecedented period of high unemployment combined with high inflation. Productivity fell dramatically. The rate of innovation also plateaued as continued technological progress in many areas became more difficult. Jet aircraft changed very little. Both Apple and Microsoft were founded during this period, but the full impact of information technology was still far in the future.

The 1980s saw increased innovation, but it became more focused in the information technology sector. This type of innovation had a different impact on workers; for those with the right skill set, computers increased their value, just as the innovations in the postwar era had done for nearly everyone. For many other workers, however, computers had a less positive effect. Some types of jobs began to be either destroyed entirely or deskilled, making workers less valuable—at least until they were able to retrain for jobs that leveraged computer technology. As information technology gained in importance, labor's share of income gradually began to decline. Jet aircraft remained largely unchanged from the 1970s but increasingly used computers in their instrumentation and controls.

The 1990s saw IT innovation accelerate even more, and the Internet took off in the second half of the decade. The trends that began in the 1980s continued, but the decade also saw the tech bubble and the creation of millions of new jobs, especially in the IT sector. These were good jobs that often involved administering the computers and networks that were rapidly becoming critical to businesses of all sizes. As a result, wages did better in this period, but still fell well short of productivity growth. Innovation was centered even more on IT. The recession of 1990–1991 was followed by a jobless recovery as workers, many of whom had lost good mid-range jobs, struggled to find new positions. The job market gradually became more polarized. Jet aircraft were still essentially similar to the designs of the 1970s; however, they now had "fly by wire" systems, in which computers moved the control surfaces in response to the pilots' inputs, as well as increased flight automation.

In the years following 2000, information technology continued its acceleration and productivity rose as businesses got better at taking full advantage of all the new innovations. Many of those good jobs created in the 1990s began to disappear as corporations automated or offshored jobs, or began to outsource their IT departments to centralized "cloud" computing services. Throughout the economy, computers and machines were increasingly replacing workers rather than making them more valuable, and wage increases fell far short of growth in productivity. Both the share of national income going to labor and the labor force participation rate declined dramatically.

The job market continued to polarize, and jobless recoveries became the norm. Jet aircraft still used the same basic designs and propulsion systems as in the 1970s, but computer-aided design and simulation had resulted in many incremental improvements in areas such as fuel efficiency. The information technology incorporated into aircraft became even more sophisticated and routinely included full-flight automation, which allowed the planes to take off, fly to a destination, and then land—all without human intervention.

Now, you may quite rightfully object to that story as being overly simplistic—or perhaps even completely wrong. After all, wasn't it really globalization, or maybe Reaganomics, that led to all our problems? As I said, this was intended to be a hypothetical narrative: a simple story to help clarify the argument for the importance of technology in these seven documented economic trends. Each of these trends has been studied by teams of economists and others who have attempted to discover the underlying causes, and technology has often been implicated as a contributing, if not always the primary, factor. However, it is when all seven trends are considered together that the argument for advancing information technology as a disruptive economic force is most compelling.

Aside from advancing information technology, there are three other primary possibilities that might conceivably have contributed to all, or at least most, of our seven economic trends: globalization, the growth of the financial sector, and politics (in which I include factors like deregulation and the decline of organized labor).

Globalization

That globalization has had a dramatic impact on certain industries and regions is undeniable—just look at America's rustbelt. But globalization, and in particular trade with China, alone could not have caused wages for most American workers to stagnate over four decades.

First, global trade directly impacts workers who are employed in the tradable sector—in other words, in industries that produce goods or services that can be transported to other locations. The vast majority of American workers now work in nontradable areas like government, education, health care, food services, and retail. For the most part, these people are not directly competing with overseas workers, so globalization is not driving down their wages.

Second, although it may appear that virtually everything sold at Walmart is made in China, most American consumer spending stays in the United States. A 2011 analysis by Galina Hale and Bart Hobijn, two economists at the Federal Reserve Bank of San Francisco, found that 82 percent of the goods and services Americans purchase are produced entirely in the United States; this is largely because we spend the vast majority of our money on nontradable services. The total value of imports from China amounted to less than 3 percent of US consumer spending.[50]

It is undoubtedly true that, as Figure 2.8 shows, the fraction of American workers employed in manufacturing has fallen

Figure 2.8. Percentage of US Workers in Manufacturing

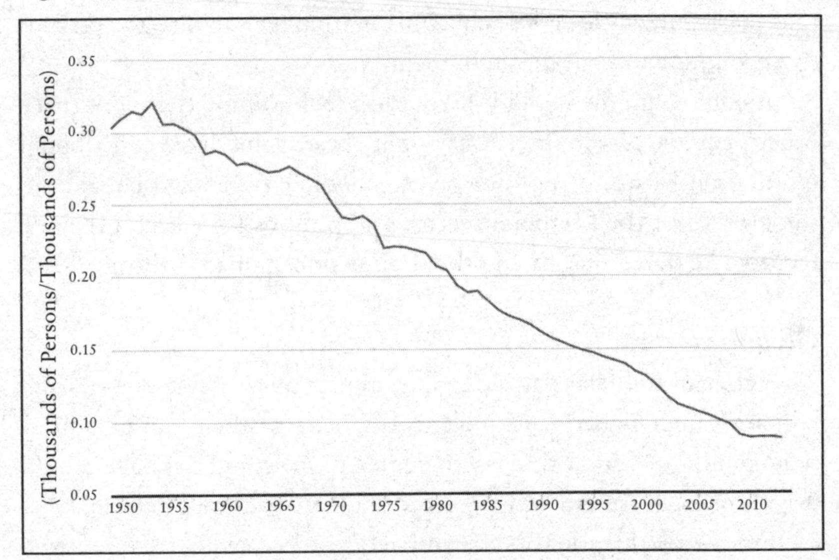

SOURCE: US Bureau of Labor Statistics and Federal Reserve Bank of St. Louis (FRED).[51]

dramatically since the early 1950s. This trend began decades before enactment of the North American Free Trade Agreement (NAFTA) in the 1990s and the rise of China in the 2000s. In fact, the decline seems to have halted at the end of the Great Recession as manufacturing employment has actually outperformed the job market as a whole.

A potent force has been very consistently eliminating jobs in the manufacturing sector. That force is advancing technology. Even as the number of manufacturing jobs has been steadily declining as a percentage of total employment, the inflation-adjusted value of the goods manufactured in the United States has dramatically increased over time. We are making more stuff, but doing so with fewer and fewer workers.

Financialization

In 1950, the US financial sector represented about 2.8 percent of the overall economy. By 2011 finance-related activity had grown more than threefold to about 8.7 percent of GDP. The compensation paid to workers in the financial sector has also exploded over the past three decades, and is now about 70 percent more than the average for other industries.[52] The assets held by banks have ballooned from about 55 percent of GDP in 1980 to 95 percent in 2000, while the profits generated in the financial sector have more than doubled from an average of about 13 percent of all corporate profits in the 1978–1997 timeframe to 30 percent in the period between 1998 and 2007.[53] No matter how you choose to measure it, finance has grown dramatically as a share of economic activity in the United States and, to a somewhat less spectacular degree, in nearly all industrialized countries.

The primary complaint leveled against the financialization of the economy is that much of this activity is geared toward rent seeking. In other words, the financial sector is not creating real value or adding to the overall welfare of society; it is simply finding ever more creative ways to siphon profits and wealth from elsewhere in the economy. Perhaps the most colorful articulation of this accusation came from *Rolling Stone*'s Matt Taibbi in his July 2009 takedown of Goldman Sachs that famously labeled the Wall Street firm "a great vampire squid wrapped around the face of humanity, relentlessly jamming its blood funnel into anything that smells like money."[54]

Economists who have studied financialization have found a strong correlation between the growth of the financial sector and inequality as well as the decline in labor's share of national income.[55] Since the financial sector is, in effect, imposing a kind of tax on the rest of the economy and then reallocating the proceeds to the top of the income distribution, it's reasonable to conclude that it has played a role in a number of the trends we've looked at. Still, it seems hard to make a strong case for financialization as the primary cause of, say, polarization and the elimination of routine jobs.

It's also important to realize that growth in the financial sector has been highly dependent on advancing information technology. Virtually all of the financial innovations that have arisen in recent decades—including, for example, collateralized debt obligations (CDOs) and exotic financial derivatives—would not have been possible without access to powerful computers. Likewise, automated trading algorithms are now responsible for nearly two-thirds of stock market trades, and Wall Street firms have built huge computing centers in close physical proximity to exchanges in order to gain trading advantages measured in tiny fractions of a second. Between 2005 and 2012, the average time to execute a trade dropped from about 10 seconds to just 0.0008 seconds,[56] and robotic, high-speed trading was heavily implicated in the May 2010 "flash crash" in which the Dow Jones Industrial Average plunged nearly a thousand points and then recovered for a net gain, all within the space of just a few minutes.

Viewed from this perspective, financialization is not so much a competing explanation for our seven economic trends; it is rather—at least to some extent—one of the ramifications of accelerating information technology. In this, there is a strong cautionary note as we look to the future: as IT continues its relentless progress, we can be certain that financial innovators, in the absence of regulations that constrain them, will find ways to leverage all those new capabilities—and, if history is any guide, it won't necessarily be in ways that benefit society as a whole.

Politics

In the 1950s, more than a third of the US private sector workforce was unionized. By 2010, that number had declined to about 7 percent.[57] At the height of its power, organized labor was a powerful advocate for the middle class as a whole. The fact that workers were able to consistently capture the lion's share of productivity growth in the 1950s and '60s can likely be attributed at least in part to the negotiating power of unions during that period. The situation today is very different; unions now struggle simply to maintain their existing membership.

The precipitous decline in the power of organized labor is one of the most visible developments associated with the rightward drift that has characterized American economic policy over the past three decades. In their 2010 book *Winner Take All Politics,* political scientists Jacob S. Hacker and Paul Pierson make a compelling case for politics as the primary driver of inequality in the United States. Hacker and Pierson point to 1978 as the pivotal year when the American political landscape began to shift under a sustained and organized assault from conservative business interests. In the decades that followed, industries were deregulated, top marginal tax rates on the wealthy and on corporations were cut to historic lows, and workplaces were made increasingly inhospitable to union organization. Much of this was driven not by electoral politics but, rather, by continuous lobbying on the part of business interests. As the power of organized labor withered, and as the number of lobbyists in Washington exploded, the day-to-day political warfare in the capital became increasingly asymmetric.

While the political situation in the United States seems uniquely detrimental to the middle class, evidence for the impact of advancing technology can be found in a wide range of developed and developing nations. Inequality is increasing in nearly all industrialized countries, while the share of national income claimed by labor is generally falling. Job market polarization has been observed in a majority of European nations. And in Canada—where organized labor remains a powerful national force—inequality is rising, median household incomes have fallen in real terms since 1980, and private sector union membership has declined as manufacturing jobs have disappeared.[58]

To some extent, the question here is one of categorization: if a nation fails to implement policies designed to mitigate the impact of structural changes brought on by advancing technology, should we label that as a problem caused by technology, or politics? Regardless, there is little question that the United States stands alone in terms of the political decisions it has made; rather than simply failing to enact policies that might have slowed the forces driving the country toward higher levels of inequality, America very often has made choices that have effectively put a wind at the back of those forces.

Looking to the Future

The debate over the primary causes of the soaring inequality and decades-long wage stagnation that have developed in the United States is likely to continue unabated, and because it touches on intensely polarizing issues—organized labor, tax rates on the wealthy, free trade, the proper role of government—the dialogue is sure to be colored by ideology. To my mind, the evidence I've presented here demonstrates that information technology has played a significant—though not necessarily dominant—role over the past few decades. Beyond that, I'm content to leave it to economic historians to delve into the data and perhaps someday shine a more definitive light on the precise forces involved in getting us to this point. The real question—and the primary subject of this book—is, What will be most important in the future? Many of the forces that heavily impacted the economy and political environment over the past half-century have largely played out. Unions outside the public sector have been decimated. Women who want careers have entered the workforce or enrolled in colleges and professional schools. There is evidence that the drive toward factory offshoring has slowed significantly, and in some cases, manufacturing is returning to the United States.

Among the forces poised to shape the future, information technology stands alone in terms of its exponential progress. Even in nations whose political environments are far more responsive to the welfare of average workers, the changes wrought by technology are becoming increasingly evident. As the technological frontier advances, many jobs that we would today consider nonroutine, and therefore protected from automation, will eventually be pulled into the routine and predictable category. The hollowed-out middle of the already polarized job market is likely to expand as robots and self-service technologies eat away at low-wage jobs, while increasingly intelligent algorithms threaten higher-skill occupations. Indeed, a 2013 study by Carl Benedikt Frey and Michael A. Osborne at the University of Oxford concluded that occupations amounting to nearly half of US total employment may be vulnerable to automation within roughly the next two decades.[59]

While accelerating information technology is nearly certain to have an outsized impact on the future economy and job market, it will remain deeply intertwined with other powerful forces. The line between technology and globalization will blur as higher-skill jobs become more vulnerable to electronic offshoring. If, as seems likely, advancing technology continues to drive the United States and other industrialized countries toward ever higher inequality, then the political influence wielded by the financial elite can only increase. This may make it even more difficult to enact policies that might serve to counteract the structural shifts occurring in the economy and improve the prospects for those in the middle and bottom of the income distribution.

In my 2009 book *The Lights in the Tunnel,* I wrote that "while technologists are actively thinking about, and writing books about, intelligent machines, the idea that technology will ever truly replace a large fraction of the human workforce and lead to permanent, structural unemployment is, for the majority of economists, almost unthinkable." To their credit, some economists have since begun to take the potential for widespread automation more seriously. In their 2011 ebook *Race Against the Machine,* Erik Brynjolfsson and Andrew McAfee of the Massachusetts Institute of Technology helped bring these ideas into the economic mainstream. Prominent economists including Paul Krugman and Jeffrey Sachs have likewise written about the possible impact of machine intelligence.[60] Nonetheless, the idea that technology might someday truly transform the job market and ultimately demand fundamental changes to both our economic system and the social contract remains either completely unacknowledged or at the very fringes of public discourse.

Indeed, among practitioners of economics and finance there is often an almost reflexive tendency to dismiss anyone who argues that this time might be different. This is very likely the correct instinct when one is discussing those aspects of the economy that are primarily driven by human behavior and market psychology. The psychological underpinnings of the recent housing bubble and bust were almost certainly little different from those that have characterized financial crises throughout history. Many of the political machinations of the early Roman republic could probably be dropped seamlessly onto the front page of today's *Politico*. These things never really change.

It would be a mistake, however, to apply that same reasoning to the impact of advancing technology. Up until the moment the first aircraft achieved sustained powered flight at Kitty Hawk, North Carolina, it was an incontrovertible fact—supported by data stretching back to the beginning of time—that human beings, strapped into heavier-than-air contraptions, *do not fly.* Just as that reality shifted in an instant, a similar phenomenon plays out continuously in nearly every sphere of technology. This time is always different where technology is concerned: that, after all, is the entire point of innovation. Ultimately, the question of whether smart machines will someday eclipse the capability of average people to perform much of the work demanded by the economy will be answered by the nature of the technology that arrives in the future—not by lessons gleaned from economic history.

IN THE NEXT CHAPTER, we'll examine the nature of information technology and its relentless acceleration, the characteristics that set it apart, and the ways in which it is already transforming important spheres of the economy.

Chapter 3

INFORMATION TECHNOLOGY: AN UNPRECEDENTED FORCE FOR DISRUPTION

Imagine depositing a penny in a bank account. Now, double the account balance every day. On day three you would go from 2 cents to 4 cents. The fifth day would take your balance from 8 to 16 cents. After less than a month, you would have more than a million dollars. If we had deposited that initial penny in 1949, just as Norbert Wiener was writing his essay about the future of computing, and then let Moore's Law run its course—doubling the amount roughly every two years—by 2015, our technological account would contain nearly $86 million. And as things move forward from this point, that balance will continue to double. Future innovations will be able to leverage that enormous accumulated balance, and as a result the rate of progress in the coming years and decades is likely to far exceed what we have become accustomed to in the past.

Moore's Law is the best-known measure of advancing computer power, but information technology is, in fact, accelerating on many different fronts. For example, computer memory capacity and the amount of digital information that can be carried on fiber-optic lines have both experienced consistent exponential increases. Nor is the acceleration confined to computer hardware; the efficiency of some software algorithms has soared at a rate far in excess of what Moore's Law alone would predict.

While exponential acceleration offers valuable insight into the advance of information technology over relatively long periods, the short-term reality is more complex. Progress is generally not always smooth and consistent; instead, it often lurches forward and then pauses while new capabilities are assimilated into organizations and the foundation for the next period of rapid advance is established. There are also intricate interdependencies and feedback loops between different realms of technology. Progress in one area may drive a sudden burst of innovation in another. As information technology marches forward, its tentacles reach ever deeper into organizations and the overall economy, often transforming the way people work in ways that can further its own advance. Consider, for example, how the rise of the Internet and sophisticated collaboration software has enabled the offshoring of software development; this has made a vastly expanded population of skilled programmers available, and all that new talent is helping to drive still more progress.

Acceleration Versus Stagnation

As information and communications technologies have advanced in their decades-long exponential march, innovation in other areas has been largely incremental. Examples include the basic design of cars, homes, aircraft, kitchen appliances, and our overall transportation and energy infrastructures, none of which, for the most part, have changed significantly since the middle of the twentieth century. PayPal co-founder Peter Thiel's famous comment—"We were promised flying cars, and instead what we got was 140 characters"—captures the sentiment of a generation that expected the future to be way cooler than this.

This lack of broad-based progress stands in stark contrast to what a person who lived through the final decades of the nineteenth century and the first half of the twentieth would have experienced. Indoor plumbing, automobiles, airplanes, electricity, home appliances, and public sanitation and utility systems all came into widespread use during this period. In industrialized countries, at least, people at all levels of society received an astonishing upgrade in the quality of their lives, even as the overall wealth of society was propelled to dizzying new heights.

Some economists have taken note of this plodding rate of advance in most spheres of technology and have tied it to the economic trends we looked at in the previous chapter, and in particular to the stagnation of incomes for most ordinary Americans. One of the foundational principles of modern economics is that such technological change is essential to long-term economic growth. Robert Solow, the economist who formalized this idea, received the Nobel Prize for his work in 1987. If innovation is the primary driver of prosperity, then perhaps stagnant incomes imply that the problem is the rate at which new inventions and ideas are being generated, rather than the impact of technology on the working and middle classes. Maybe computers aren't really all that important, and the slow rate of progress on a broader front is what matters most.

Several economists have made this case. Tyler Cowen, an economist at George Mason University, proposed in his 2011 book *The Great Stagnation* that the US economy has run into a temporary plateau after consuming all the low-hanging fruit of accessible innovation, free land, and underutilized human talent. Robert J. Gordon of Northwestern University is even more pessimistic, arguing in a 2012 paper that economic growth in the United States, hampered by a slow pace of innovation and a number of "headwinds"—including excessive debt, an aging population, and shortfalls in our educational system—may essentially be over.[1]

In order to gain some insight into the factors that influence the pace of innovation, we may find it useful to think in terms of the historical path that nearly all technologies follow. Airplanes are a good example. The first controlled, powered flight occurred in December 1903 and lasted about twelve seconds. Progress accelerated from that humble start, but the primitive initial level of the technology meant it would take years before a practical airplane would emerge. By 1905, Wilbur Wright was able to stay aloft for nearly forty minutes while traveling about twenty-four miles. Within a few years, however, things started to really come together; aircraft technology had progressed along its exponential curve, and the rate of absolute progress picked up dramatically. By World War I, airplanes were engaging in high-speed aerial dog fights. Progress continued its acceleration over the next two decades, ultimately producing high-performance fighter aircraft like the Spitfire, the Zero, and the P-51. Sometime around World War II, however, the rate of advance slowed significantly. Aircraft powered by internal combustion engines driving propellers were now very close to their ultimate technical potential, and design improvements beyond that point would be incremental.

This S-shaped path in which accelerating—or exponential—advance ultimately matures into a plateau effectively illustrates the life story of virtually all specific technologies. Of course, we know that as World War II came to a close, an entirely new aircraft technology appeared on the scene. Jet aircraft would soon offer a level of performance far beyond what was possible for any propeller-driven plane. Jets were a disruptive technology: they had an S-curve of their own. Figure 3.1 shows what this might look like.

If we want to dramatically speed up the pace of innovation in aircraft design, we need to find yet another S-curve, and that curve has to represent a technology that is not only superior in terms of performance but also economically viable.* The problem, of course,

* The supersonic Concorde, for example, offered a new S-curve in terms of absolute performance, but it did not prove to be an economically sustainable technology and was never able to capture more than a tiny fraction of the airline passenger market. The Concorde was in service from 1976 until 2003.

Figure 3.1. Aircraft Technology S-Curves

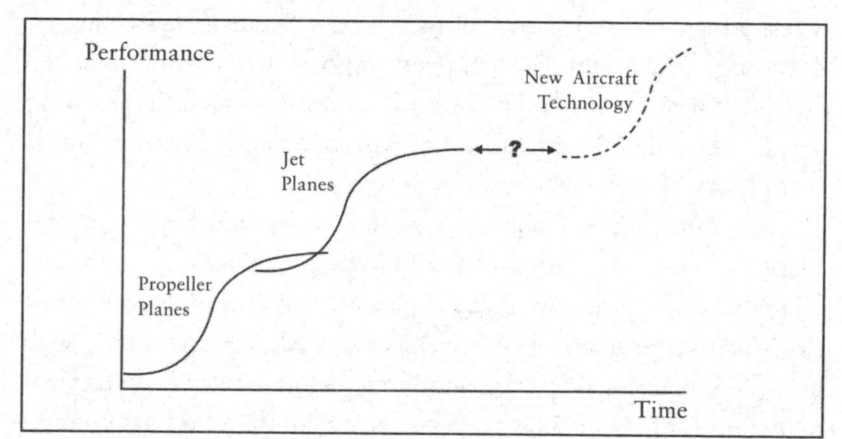

is that so far, that new curve is nowhere to be found. Assuming we can't discover this disruptive new technology simply by hopping the fence at Area 51, it's going to take a giant leap to get to that new S-curve—and this presumes, of course, that the curve even exists.

The critical point here is that while many factors, such as the level of research and development effort and investment, or the presence of a favorable regulatory environment, can certainly have an impact on the relative position of technology S-curves, the most important factor by far is the set of physical laws that govern the sphere of technology in question. We don't yet have a disruptive new aircraft technology and that is primarily due to the laws of physics and the limitations they imply relative to our current scientific and technical knowledge. If we hope to have another period of rapid innovation in a wide range of technological areas—perhaps something comparable to what occurred between approximately 1870 and 1960—we would need to find new S-curves in all these different areas. Obviously, that is likely to represent an enormous challenge.

There is one important reason for optimism, however, and that is the positive impact that accelerating information technology will have on research and development in other fields. Computers have already been transformative in many areas. Sequencing the human genome would certainly have been impossible without advanced computing power. Simulation and computer-based design have greatly expanded the potential for experimentation with new ideas in a variety of research areas.

One information technology success story that has had a dramatic and personal impact on all of us has been the role of advanced computing power in oil and gas exploration. As the global supply of easily accessible oil and gas fields has declined, new techniques such as three-dimensional underground imaging have become indispensable tools for locating new reserves. Aramco, the Saudi national oil company, for example, maintains a massive computing center where powerful supercomputers are instrumental in maintaining the flow of oil. Many people might be surprised to learn that one of the most important ramifications of Moore's Law has been the fact that, at least so far, world energy supplies have kept pace with surging demand.

The advent of the microprocessor has resulted in an astonishing increase in our overall ability to perform computations and manipulate information. Where once computers were massive, slow, expensive, and few in number, today they are cheap, powerful, and ubiquitous. If you were to multiply a single computer's increase in computational power since 1960 by the number of new microprocessors that have appeared since then, the result would be nearly beyond reckoning. It seems impossible to imagine that such an immeasurable increase in our overall computing capacity won't eventually have dramatic consequences in a variety of scientific and technical fields. Nonetheless, the primary determinant of the positions of the technology S-curves we'll need to reach in order to have truly disruptive innovation is still the applicable laws of nature. Computational capability can't change that reality, but it may well help researchers to bridge some of the gaps.

The economists who believe we have hit a technological plateau typically have deep faith in the relationship between the pace of innovation and the realization of broad-based prosperity; the implication is that if we can just jump-start technological progress on a broad front, median incomes will once again begin increasing in real terms. I think there are good reasons to be concerned that this may not necessarily turn out to be the case. In order to understand why, let's look at what makes information technology unique and the ways in which it will intertwine with innovations in other areas.

Why Information Technology Is Different

The relentless acceleration of computer hardware over decades suggests that we've somehow managed to remain on the steep part of the S-curve for far longer than has been possible in other spheres of technology. The reality, however, is that Moore's Law has involved successfully climbing a staircase of cascading S-curves, each representing a specific semiconductor fabrication technology. For example, the lithographic process used to lay out integrated circuits was initially based on optical imaging techniques. When the size of individual device elements shrank to the point where the wavelength of visible light was too long to allow for further progress, the semiconductor industry moved on to X-ray lithography.[2] Figure 3.2 illustrates roughly what climbing a series of S-curves might look like.

One of the defining characteristics of information technology has been the relative accessibility of subsequent S-curves. The key to sustainable acceleration has not been so much that the fruit is low-hanging but, rather, that the tree is climbable. Climbing that tree has been a complex process that has been driven by intensive competition and has required enormous investment. There has also been substantial cooperation and planning. To help coordinate all these efforts, the industry publishes a massive document called the International Technology Roadmap for Semiconductors (ITRS), which essentially offers a detailed fifteen-year preview of how Moore's Law is expected to unfold.

Figure 3.2. Moore's Law as a Staircase of S-Curves

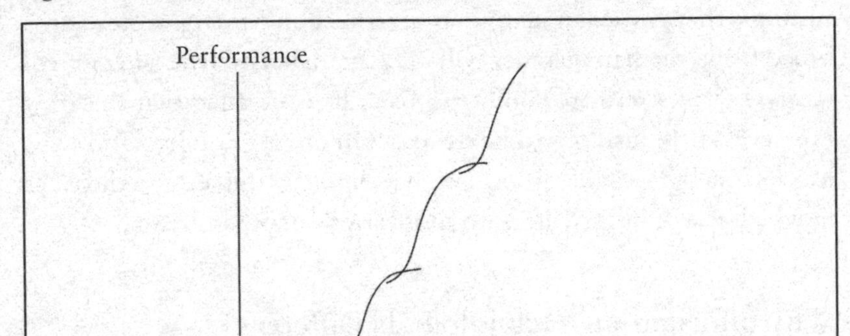

As things stand today, computer hardware may soon run into the same type of challenge that characterizes other areas of technology. In other words, reaching that next S-curve may eventually require a giant—and perhaps even unachievable—leap. The historical path followed by Moore's Law has been to keep shrinking the size of transistors so that more and more circuitry can be packed onto a chip. By the early 2020s, the size of individual design elements on computer chips will be reduced to about five nanometers (billionths of a meter), and that is likely to be very close to the fundamental limit beyond which no further miniaturization is possible. There are, however, a number of alternate strategies that may allow progress to continue unabated, including three-dimensional chip design and exotic carbon-based materials.[3] *

* The idea behind 3D chips is to begin stacking circuitry vertically in multiple layers. Samsung Electronics began manufacturing 3D flash memory chips in August 2013. If this technique proves economically viable for the far more sophisticated processor chips designed by companies like Intel and AMD (Advanced Micro Devices), it may represent the future of Moore's Law. Another possibility is to turn to exotic carbon-based materials as an alternative to silicon. Graphene and carbon nanotubes, both of which are the result of recent nanotechnology research, may eventually offer a new medium for very high-performance computing. Researchers at Stanford University have already created a rudimentary carbon nanotube computer, although its performance falls far short of commercial silicon-based processors.

Even if the advance of computer hardware capability were to plateau, there would remain a whole range of paths along which progress could continue. Information technology exists at the intersection of two different realities. Moore's Law has dominated the realm of atoms, where innovation is a struggle to build faster devices and to minimize or find a way to dissipate the heat they generate. In contrast, the realm of bits is an abstract, frictionless place where algorithms, architecture (the conceptual design of computing systems), and applied mathematics govern the rate of progress. In some areas, algorithms have already advanced at a far faster rate than hardware. In a recent analysis, Martin Grötschel of the Zuse Institute in Berlin found that, using the computers and software that existed in 1982, it would have taken a full eighty-two years to solve a particularly complex production planning problem. As of 2003, the same problem could be solved in about a minute—an improvement by a factor of around 43 million. Computer hardware became about 1,000 times faster over the same period, which means that improvements in the algorithms used accounted for approximately a 43,000-fold increase in performance.[4]

Not all software has improved so quickly. This is especially true of areas where software must interact directly with people. In an August 2013 interview with James Fallows of *The Atlantic,* Charles Simonyi, the computer scientist who oversaw the development of Microsoft Word and Excel, expressed the view that software has largely failed to leverage the advances that have occurred in hardware. When asked where the most potential for future improvement lies, Simonyi said: "The basic answer is that nobody would be doing routine, repetitive things anymore."[5]

There is also tremendous room for future progress through finding improved ways to interconnect vast numbers of inexpensive processors in massively parallel systems. Reworking current hardware device technology into entirely new theoretical designs could likewise produce giant leaps in computer power. Clear evidence that a sophisticated architectural design based on deeply complex interconnection can produce astonishing computational capability is provided by what is, by far, the most powerful general computing machine in existence: the human brain. In creating the brain, evolution did not have the luxury of Moore's Law. The "hardware" of a human brain is no faster than that of a mouse and is thousands to millions of times slower than a modern integrated circuit; the difference lies entirely in the sophistication of the design.[6] Indeed, the ultimate in computer capability—and perhaps machine intelligence—might be achieved if someday researchers are able to marry the speed of even today's computer hardware with something approaching the level of design complexity you would find in the brain. Baby steps have already been taken in that direction: IBM released a cognitive computing chip—inspired by the human brain and aptly branded "SyNAPSE"—in 2011 and has since created a new programming language to accompany the hardware.[7]

Beyond the relentless acceleration of hardware, and in many cases software, there are, I think, two other defining characteristics of information technology. The first is that IT has evolved into a true general-purpose technology. There are very few aspects of our daily lives, and especially of the operation of businesses and organizations of all sizes, that are not significantly influenced by or even highly dependent on information technology. Computers, networks, and the Internet are now irretrievably integrated into our economic, social, and financial systems. IT is everywhere, and it's difficult to even imagine life without it.

Many observers have compared information technology to electricity, the other transformative general-purpose technology that came into widespread use in the first half of the twentieth century. Nicholas Carr makes an especially compelling argument for viewing IT as an electricity-like utility in his 2008 book *The Big Switch*. While many of these comparisons are apt, the truth is that electricity is a tough act to follow. Electrification transformed businesses, the overall economy, social institutions, and individual lives to an astonishing degree—and it did so in ways that were overwhelmingly positive. It would probably be very difficult to find a single person in a developed country like the United States who did not eventually receive a major upgrade in his or her standard of living after the advent of electric power. The transformative impact of information technology is likely to be more nuanced and, for many people, less universally positive. The reason has to do with IT's other signature characteristic: cognitive capability.

Information technology, to a degree that is unprecedented in the history of technological progress, encapsulates intelligence. Computers make decisions and solve problems. Computers are machines that can—in a very limited and specialized sense—*think*. No one would argue that today's computers approach anything like human-level general intelligence. But that very often misses the point. Computers are getting dramatically better at performing specialized, routine, and predictable tasks, and it seems very likely that they will soon be poised to outperform many of the people now employed to do these things.

Progress in the human economy has resulted largely from occupational specialization, or as Adam Smith would say, "the division of labour." One of the paradoxes of progress in the computer age is that as work becomes ever more specialized, it may, in many cases, also become more susceptible to automation. Many experts would say that, in terms of *general* intelligence, today's best technology barely outperforms an insect. And yet, insects do not make a habit of landing jet aircraft, booking dinner reservations, or trading on Wall Street. Computers now do all these things, and they will soon begin to aggressively encroach in a great many other areas.

Comparative Advantage and Smart Machines

Economists who reject the idea that machines could someday make a large fraction of our workforce essentially unemployable often base their argument on one of the biggest ideas in economics: the theory of comparative advantage.[8] To see how comparative advantage works, let's consider two people. Jane is truly exceptional. After many years of intensive training and a record of nearly unmatched success, she is considered to be one of the world's leading neurosurgeons. In her gap years between college and medical school, Jane enrolled in one of France's best culinary institutes and is now also a gourmet cook of rarefied talent. Tom is more of an average guy. He is, however, a very good cook, and has been complimented many times on his skills. Still, he can't really come close to matching what Jane can do in the kitchen. And it goes without saying that Tom wouldn't be allowed anywhere near an operating room.

Given that Tom can't compete with Jane as a cook, and certainly not as a surgeon, is there any way that the two could enter into an agreement that would make them both better off? Comparative advantage says "yes" and tells us that Jane could hire Tom as a cook. Why would she do that when she can get a better result by doing the cooking herself? The answer is that it would free up more of Jane's time and energy for the one thing she is truly exceptional at (and the thing that brings in the most income): brain surgery.

The main idea behind comparative advantage is that you should always be able to find a job, provided you specialize in the thing at which you are "least bad" relative to other people. By doing so, you offer others the chance to also specialize and thereby earn a higher income. In Tom's case, least bad meant cooking. Jane is luckier (and a lot richer) because her least bad gig is something she is truly great at, and that talent happens to have a very high market value. Throughout economic history, comparative advantage has been the primary driver of ever more specialization and trade between individuals and nations.

Now let's change the story. Imagine that Jane has the ability to easily and inexpensively clone herself. If you like science fiction movies, think in terms of *Matrix Reloaded*, where Neo battles dozens of copies of the agent, Smith. In that particular struggle, Neo ultimately prevails, but I think you can see that Tom might not be so lucky when it comes to keeping his job working for Jane. Comparative advantage works because of opportunity cost: if a person chooses to do one thing, she must necessarily give up the opportunity to do something else. Time and space are limited; she can't be in two places doing two things at once.

Machines, and particularly software applications, can be easily replicated. In many cases they can be cloned at a cost that is small compared with employing a person. When intelligence can be replicated, the concept of opportunity cost is upended. Jane can now perform brain surgery and cook simultaneously. So why does she need Tom at all? It's a good bet that pretty soon Jane's clones will also start putting less talented brain surgeons out of work. Comparative advantage in the age of smart machines might require something of a rethink.

Imagine the impact of a large corporation being able to train a single employee and then clone him into an army of workers, all of whom instantly possess his knowledge and experience but, from that point on, are also capable of continuing to learn and adapt to new situations. When the intelligence encapsulated in information technology is replicated and scaled across organizations, it has the potential to fundamentally redefine the relationship between people and machines. From the perspective of a great many workers, computers will cease to be tools that enhance their productivity and instead become viable substitutes. This outcome will, of course, dramatically increase the productivity of many businesses and industries—but it will also make them far less labor-intensive.

The Tyranny of the Long Tail

The influence of this distributed machine intelligence is most evident in the information technology industry itself. The Internet has spawned enormously profitable and influential corporations with startlingly diminutive workforces. In 2012, Google, for example, generated a profit of nearly $14 billion while employing fewer than 38,000 people.[9] Contrast that with the automotive industry. At peak employment in 1979, General Motors alone had nearly 840,000 workers but earned only about $11 billion—20 percent less than what Google raked in. And, yes, that's after adjusting for inflation.[10] Ford, Chrysler, and American Motors employed hundreds of thousands more people. Beyond that core workforce, the industry also created millions of peripheral middle-class jobs in areas like driving, repairing, insuring, and renting cars.

Of course, the Internet sector also offers peripheral opportunities. The new information economy is often touted as the great equalizer. After all, anyone can write a blog and run ads on it, publish an ebook, sell stuff on eBay, or develop an iPhone app. While these opportunities do indeed exist, they are dramatically different from all those solid middle-class jobs created by the automotive industry. The evidence shows pretty clearly that the income realized from online activities nearly always tends to follow a winner-take-all distribution. While the Internet may, in theory, equalize opportunity and demolish entry barriers, the actual outcomes it produces are almost invariably highly unequal.

If you graph the traffic coming to websites, advertising revenue generated online, music downloads from the iTunes store, books sold on Amazon, apps downloaded from Apple's AppStore or Google Play, or just about anything else online, you will nearly always end up with something that looks like Figure 3.3. This ubiquitous long-tail distribution is central to the business models of the corporations that dominate the Internet sector. Companies like Google, eBay, and Amazon are able to generate revenue from *every point* on the distribution. If a company controls a large market, then aggregating even tiny sums along the entire curve results in total revenues that can easily reach into the billions.

Figure 3.3. A Winner-Take-All/Long-Tail Distribution

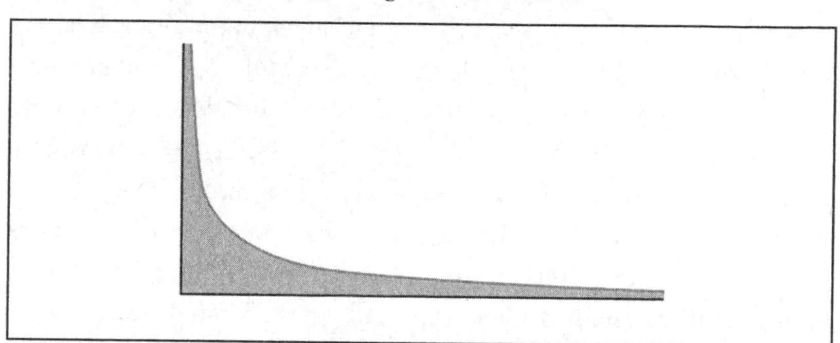

Markets in goods and services that are susceptible to digitalization inevitably evolve into this winner-take-all distribution. Sales of books and music, classified advertising, and movie rentals, for example, are increasingly dominated by a tiny number of online distribution hubs, and one obvious result has been the elimination of vast numbers of jobs for people like journalists and retail store clerks.

The long tail is great if you own it. When, however, you occupy only a single point on the distribution, the story is quite different. Out on the long tail, incomes from most online activities rapidly drop to the pocket-change level. That can work out fine if you have an alternate source of income, or if you happen to be living in your parents' basement. The problem is that as digital technology continues to transform industries, more and more of the jobs that provide that primary-income source are likely to disappear.

As more people lose the dependable income stream that anchors them into the middle class, they are likely to increasingly turn to these long-tail opportunities in the digital economy. A lucky few will provide the anecdotal success stories we will hear about, but the vast majority will struggle to maintain anything approaching a middle-class lifestyle. As techno-visionary Jaron Lanier has pointed out, a great many people are likely to be forced into the type of informal economy that is found in third-world nations.[11] Young adults who find the freedom of the informal economy alluring will quickly discover its drawbacks when they begin to think in terms of maintaining a home, raising children, or planning for retirement. Of course, there have always been people living at the fringes in the United States and other developed economies, but to some extent they free-ride on the wealth generated by a critical mass of middle-class households. The presence of that solid middle is one of the primary factors that differentiates an advanced nation from an impoverished one—and its erosion is becoming increasingly evident, especially in the United States.

Most techno-optimists would likely object to this characterization. They tend to view information technology as universally empowering. It is perhaps not coincidental that they also tend to have been very successful in the new economy. The most prominent digital optimists typically live at the extreme left of the long tail—or, even better, they've perhaps founded a company that owns the entire distribution. In a PBS television special that aired in 2012, inventor and futurist Ray Kurzweil was asked about the possibility of a "digital divide"—meaning that only a small percentage of the population will be able to thrive in the new information economy. Kurzweil dismissed the idea of such a divide and instead pointed to empowering technologies like mobile phones. Anybody with a smart phone, he said, "is carrying around billions of dollars of capability circa 20 or 30 years ago."[12] Left unsaid was how the average person is supposed to leverage that technology into a livable income.

Mobile phones have indeed been shown to improve living standards, but this has been documented primarily in developing countries that lack other communications infrastructure. By far the most celebrated success story involves sardine fishermen in Kerala, a region along the southwest coast of India. In a 2007 research paper, economist Robert Jensen described how mobile phones allowed the fishermen to determine which villages offered the best markets for their fish.[13] Before the advent of wireless technology, targeting a particular village was a guess that often resulted in a mismatch between supply and demand. However, with their new phones, the fishermen knew exactly where the buyers were, and this has resulted in a better functioning market with more stable prices and far less waste.

The sardine fishermen of Kerala have become a kind of standard-bearer for techno-optimism as it relates to developing countries, and their story has been told in numerous books and magazine articles.[14] While mobile phones are unquestionably of great value to third-world fishermen, there is little evidence to suggest that average citizens in developed countries—or, for that matter, even in poor countries—will succeed in deriving a meaningful income from their smart phones. Even skilled software developers find it extremely challenging to generate significant revenue from mobile apps, and the primary reason, needless to say, is that ubiquitous long-tail distribution. Visit almost any online forum populated by Android or iPhone developers and you're likely to find discussions lamenting the winner-take-all nature of the mobile ecosystem and the difficulty in monetizing apps. As a practical matter, for the majority of people who lose middle-class jobs, access to a smart phone may offer little beyond the ability to play Angry Birds while waiting in the unemployment line.

A Moral Question

If we think again in terms of doubling a penny as a proxy for the exponential advance of digital technology, it's clear that today's enormous technological account balance results from the efforts of countless individuals and organizations over the course of decades. Indeed, the arc of progress can be traced back in time at least as far as Charles Babbage's mechanical difference engine in the early nineteenth century.

The innovations that have resulted in fantastic wealth and influence in today's information economy, while certainly significant, do not really compare in importance to the groundbreaking work done by pioneers like Alan Turing or John von Neumann. The difference is that even incremental advances are now able to leverage that extraordinary accumulated account balance. In a sense, the successful innovators of today are a bit like the Boston Marathon runner who in 1980 famously snuck into the race only half a mile from the finish line.

Of course, all innovators stand on the shoulders of those who came before them. This was certainly true when Henry Ford introduced the Model T. However, as we have seen, information technology is fundamentally different. IT's unique ability to scale machine intelligence across organizations in ways that will substitute for workers and its propensity to everywhere create winner-take-all scenarios will have dramatic implications for both the economy and society.

At some point, we may need to ask a fundamental moral question: Should the population at large have some sort of claim on that accumulated technological account balance? The public does, of course, benefit greatly from accelerating digital technology in terms of lower costs, convenience, and free access to information and entertainment. But that brings us back to the problem with Kurzweil's argument about mobile phones: those things won't pay the rent.

It should be kept in mind, as well, that much of the basic research that enabled progress in the IT sector was funded by American taxpayers. The Defense Advanced Research Projects Agency (DARPA) created and funded the computer network that ultimately evolved into the Internet.* Moore's Law has come about, in part, because of university-led research funded by the National Science Foundation. The Semiconductor Industry Association, the industry's political action committee, actively lobbies for increased federal research dollars. Today's computer technology exists in some measure because millions of middle-class taxpayers supported federal funding

* DARPA also provided the initial financial backing for the development of Siri (now Apple's virtual assistant technology) and has underwritten the development of IBM's new SyNAPSE cognitive computing chips.

for basic research in the decades following World War II. We can be reasonably certain that those taxpayers offered their support in the expectation that the fruits of that research would create a more prosperous future for their children and grandchildren. Yet, the trends we looked at in the last chapter suggest we are headed toward a very different outcome.

BEYOND THE BASIC MORAL QUESTION of whether a tiny elite should be able to, in effect, capture ownership of society's accumulated technological capital, there are also practical issues regarding the overall health of an economy in which income inequality becomes too extreme. Continued progress depends on a vibrant market for future innovations—and that, in turn, requires a reasonable distribution of purchasing power.

In later chapters, we'll look in more detail at some of the overall economic and social implications of digital technology's relentless acceleration. But first, let's look at how these innovations are increasingly threatening the high-skill jobs held by workers with college and even graduate or professional degrees.

Chapter 4

WHITE-COLLAR JOBS AT RISK

On October 11, 2009, the Los Angeles Angels prevailed over the Boston Red Socks in the American League play-offs and earned the right to face the New York Yankees for the league championship and entry into the World Series. It was an especially emotional win for the Angels because just six months earlier one of their most promising players, pitcher Nick Adenhart, had been killed by a drunk driver in an automobile accident. One sportswriter began an article describing the game like this:

> Things looked bleak for the Angels when they trailed by two runs in the ninth inning, but Los Angeles recovered thanks to a key single from Vladimir Guerrero to pull out a 7–6 victory over the Boston Red Sox at Fenway Park on Sunday.
>
> Guerrero drove in two Angels runners. He went 2–4 at the plate.
>
> "When it comes down to honoring Nick Adenhart, and what happened in April in Anaheim, yes, it probably was the biggest hit [of my career]," Guerrero said. "Because I'm dedicating that to a former teammate, a guy that passed away."

> Guerrero has been good at the plate all season, especially in day games. During day games Guerrero has a .794 OPS [on-base plus slugging]. He has hit five home runs and driven in 13 runners in 26 games in day games.[1]

The author of that text is probably in no immediate danger of receiving any awards for his writing. The narrative is nonetheless a remarkable achievement: not because it is readable, grammatically correct, and an accurate description of the baseball game, but because the author is a computer program.

The software in question, called "StatsMonkey," was created by students and researchers at Northwestern University's Intelligent Information Laboratory. StatsMonkey is designed to automate sports reporting by transforming objective data about a particular game into a compelling narrative. The system goes beyond simply listing facts; rather, it writes a story that incorporates the same essential attributes that a sports journalist would want to include. StatsMonkey performs a statistical analysis to discern the notable events that occurred during a game; it then generates natural language text that summarizes the game's overall dynamic while focusing on the most important plays and the key players who contributed to the story.

In 2010, the Northwestern University researchers who oversaw the team of computer science and journalism students who worked on StatsMonkey raised venture capital and founded a new company, Narrative Science, Inc., to commercialize the technology. The company hired a team of top computer scientists and engineers; then it tossed out the original StatsMonkey computer code and built a far more powerful and comprehensive artificial intelligence engine that it named "Quill."

Narrative Science's technology is used by top media outlets, including *Forbes*, to produce automated articles in a variety of areas, including sports, business, and politics. The company's software generates a news story approximately every thirty seconds, and many of these are published on widely known websites that prefer not to acknowledge their use of the service. At a 2011 industry conference, *Wired* writer Steven Levy prodded Narrative Science co-founder Kristian Hammond into predicting the percentage of news articles that would be written algorithmically within fifteen years. His answer: over 90 percent.[2]

Narrative Science has its sights set on far more than just the news industry. Quill is designed to be a general-purpose analytical and narrative-writing engine, capable of producing high-quality reports for both internal and external consumption across a range of industries. Quill begins by collecting data from a variety of sources, including transaction databases, financial and sales reporting systems, websites, and even social media. It then performs an analysis designed to tease out the most important and interesting facts and insights. Finally, it weaves all this information into a coherent narrative that the company claims measures up to the efforts of the best human analysts. Once it's configured, the Quill system can generate business reports nearly instantaneously and deliver them continuously—all without human intervention.[3] One of Narrative Science's earliest backers was In-Q-Tel, the venture capital arm of the Central Intelligence Agency, and the company's tools will likely be used to automatically transform the torrents of raw data collected by the US intelligence community into an easily understandable narrative format.

The Quill technology showcases the extent to which tasks that were once the exclusive province of skilled, college-educated professionals are vulnerable to automation. Knowledge-based work, of course, typically calls upon a wide range of capabilities. Among other things, an analyst may need to know how to retrieve information from a variety of systems, perform statistical or financial modeling, and then write understandable reports and presentations. Writing—which, after all, is at least as much art as it is science—might seem like one of the least likely tasks to be automated. Nevertheless, it has been, and the algorithms are improving rapidly. Indeed, because knowledge-based jobs can be automated using only software, these positions may, in many cases, prove to be more vulnerable than lower-skill jobs that involve physical manipulation.

Writing also happens to be an area in which employers consistently complain that college graduates are deficient. One recent survey of employers found that about half of newly hired two-year college graduates and over a quarter of those with four-year degrees were found to have poor writing—and in some cases even reading—skills.[4] If intelligent software can, as Narrative Science claims, begin to rival the most capable human analysts, the future growth of knowledge-based employment is in doubt for all college graduates, especially the least prepared.

Big Data and Machine Learning

The Quill narrative-writing engine is just one of many new software applications being developed to leverage the enormous amounts of data now being collected and stored within businesses, organizations, and governments across the global economy. By one estimate, the total amount of data stored globally is now measured in thousands of exabytes (an exabyte is equal to a billion gigabytes), and that figure is subject to its own Moore's Law–like acceleration, doubling roughly every three years.[5] Nearly all of that data is now stored in digital format and is therefore accessible to direct manipulation by computers. Google's servers alone handle about 24 petabytes (equal to a million gigabytes)—primarily information about what its millions of users are searching for—each and every day.[6]

All this data arrives from a multitude of different sources. On the Internet alone, there are website visits, search queries, emails, social media interactions, and advertising clicks, to name just a few examples. Within businesses, there are transactions, customer contacts, internal communications, and data captured in financial, accounting, and marketing systems. Out in the real world, sensors continuously capture real-time operational data in factories, hospitals, automobiles, aircraft, and countless other consumer devices and industrial machines.

The vast majority of this data is what a computer scientist would call "unstructured." In other words, it is captured in a variety of formats that can often be difficult to match up or compare. This is very different from traditional relational database systems where information is arranged neatly in consistent rows and columns that make search and retrieval fast, reliable, and precise. The unstructured nature of big data has led to the development of new tools specifically geared toward making sense of information that is collected from a variety of sources. Rapid improvement in this area is just one more example of the way in which computers are, at least in a limited sense, beginning to encroach on capabilities that were once exclusive to human beings. The ability to continuously process a stream of unstructured information from sources throughout our environment is, after all, one of the things for which humans are uniquely adapted. The difference, of course, is that in the realm of big data, computers are able to do this on a scale that, for a person, would be impossible. Big data is having a revolutionary impact in a wide range of areas including business, politics, medicine, and nearly every field of natural and social science.

Major retailers are relying on big data to get an unprecedented level of insight into the buying preferences of individual shoppers, allowing them to make precisely targeted offers that increase revenue while helping to build customer loyalty. Police departments across the globe are turning to algorithmic analysis to predict the times and locations where crimes are most likely to occur and then deploying their forces accordingly. The City of Chicago's data portal allows residents to see both historical trends and real-time data in a range of areas that capture the ebb and flow of life in a major city—including energy usage, crime, performance metrics for transportation, schools and health care, and even the number of potholes patched in a given period of time. Tools that provide new ways to visualize data collected from social media interactions as well as sensors built into doors, turnstiles, and escalators offer urban planners and city managers graphic representations of the way people move, work, and interact in urban environments, a development that may lead directly to more efficient and livable cities.

There is a potential dark side, however. Target, Inc., provided a far more controversial example of the ways in which vast quantities of extraordinarily detailed customer data can be leveraged. A data scientist working for the company found a complex set of correlations involving the purchase of about twenty-five different health and cosmetic products that were a powerful early predictor of pregnancy. The company's analysis could even estimate a woman's due date with a high degree of accuracy. Target began bombarding women with offers for pregnancy-related products at such an early stage that, in some cases, the women had often not yet shared the news with their immediate families. In an article published in early 2012, the *New York Times* reported one case in which the father of a teenage girl actually complained to store management about mail sent to the family's home—only to find out later that Target, in fact, knew more than he did.[7] Some critics fear that this rather creepy story is only the beginning and that big data will increasingly be used to generate predictions that potentially violate privacy and perhaps even freedom.

The insights gleaned from big data typically arise entirely from correlation and say nothing about the causes of the phenomenon being studied. An algorithm may find that if A is true, B is likely also true. But it cannot say whether A causes B or vice versa—or if perhaps both A and B are caused by some external factor. In many cases, however, and especially in the realm of business where the ultimate measure of success is profitability and efficiency rather than deep understanding, correlation alone can have extraordinary value. Big data can offer management an unprecedented level of insight into a wide range of areas: everything from the operation of a single machine to the overall performance of a multinational corporation can potentially be analyzed at a level of detail that would have been impossible previously.

The ever-growing mountain of data is increasingly viewed as a resource that can be mined for value—both now and in the future. Just as extractive industries like oil and gas continuously benefit from technical advances, it's a good bet that accelerating computer power and improved software and analysis techniques will enable corporations to unearth new insights that lead directly to increased profitability. Indeed, that expectation on the part of investors is probably what gives data-intensive companies like Facebook such enormous valuations.

Machine learning—a technique in which a computer churns through data and, in effect, writes its own program based on the statistical relationships it discovers—is one of the most effective means of extracting all that value. Machine learning generally involves two steps: an algorithm is first trained on known data and is then unleashed to solve similar problems with new information. One ubiquitous use of machine learning is in email spam filters. The algorithm might be trained by processing millions of emails that have been pre-categorized as either spam or not. No one sits down and directly programs the system to recognize every conceivable typographic butchery of the word "Viagra." Instead, the software figures this out by itself. The result is an application that can automatically identify the vast majority of junk email and can also continuously improve and adapt over time as more examples become available. Machine learning algorithms based on the same basic principles recommend books at Amazon.com, movies at Netflix, and potential dates at Match.com.

One of the most dramatic demonstrations of the power of machine learning came when Google introduced its online language translation tool. Its algorithms used what might be called a "Rosetta

Stone" approach to the problem by analyzing and comparing millions of pages of text that had already been translated into multiple languages. Google's development team began by focusing on official documents prepared by the United Nations and then extended their effort to the Web, where the company's search engine was able to locate a multitude of examples that became fodder for their voracious self-learning algorithms. The sheer number of documents used to train the system dwarfed anything that had come before. Franz Och, the computer scientist who led the effort, noted that the team had built "very, very large language models, much larger than anyone has ever built in the history of mankind."[8]

In 2005, Google entered its system in the annual machine translation competition held by the National Bureau of Standards and Technology, an agency within the US Commerce department that publishes measurement standards. Google's machine learning algorithms were able to easily outperform the competition—which typically employed language and linguistic experts who attempted to actively program their translation systems to wade through the mire of conflicting and inconsistent grammatical rules that characterize languages. The essential lesson here is that, when datasets are large enough, the knowledge encapsulated in all that data will often trump the efforts of even the best programmers. Google's system is not yet competitive with the efforts of skilled human translators, but it offers bidirectional translation between more than five hundred language pairs. That represents a genuinely disruptive advance in communication capability: for the first time in human history, nearly anyone can freely and instantly obtain a rough translation of virtually any document in any language.

While there are a number of different approaches to machine learning, one of the most powerful, and fascinating, techniques involves the use of artificial neural networks—or systems that are designed using the same fundamental operating principles as the human brain. The brain contains as many as 100 billion neuron cells—and many trillions of connections between them—but it's possible to build powerful learning systems using far more rudimentary configurations of simulated neurons.

An individual neuron operates somewhat like the plastic pop-up toys that are popular with very young children. When the child pushes the button, a colorful figure pops up—perhaps a cartoon character or an animal. Press the button gently and nothing happens. Press it a bit harder and still nothing. But exceed a certain force threshold, and up pops the figure. A neuron works in essentially the same fashion, except that the activation button can be pressed by a combination of multiple inputs.

To visualize a neural network, imagine a Rube Goldberg–like machine in which a number of these pop-up toys are arranged on the floor in rows. Three mechanical fingers are poised over each toy's activation button. Rather than having a figure pop up, the toys are configured so that when a toy is activated it causes several of the mechanical fingers in the next row of toys to press down on their own buttons. The key to the neural network's ability to learn is that the force with which each finger presses down on its respective button can be adjusted.

To train the neural network, you feed known data into the first row of neurons. For example, imagine inputting visual images of handwritten letters. The input data causes some of the mechanical fingers to press down with varying force depending on their calibration. That, in turn, causes some of the neurons to activate and press down on buttons in the next row. The output—or answer—is gathered from the last row of neurons. In this case, the output will be a binary code identifying the letter of the alphabet that corresponds to the input image. Initially, the answer will be wrong, but our machine also includes a comparison and feedback mechanism. The output is compared to the known correct answer, and this automatically results in adjustments to the mechanical fingers in each row, and that, in turn, alters the sequence of activating neurons. As the network is trained with thousands of known images, and then the force with which the fingers press down is continuously recalibrated, the network will get better and better at producing the correct answer. When things reach the point where the answers are no longer improving, the network has effectively been trained.

This is, in essence, the way that neural networks can be used to recognize images or spoken words, translate languages, or perform a variety of other tasks. The result is a program—essentially a list of all the final calibrations for the mechanical fingers poised over the neuron activation buttons—that can be used to configure new neural networks, all capable of automatically generating answers from new data.

Artificial neural networks were first conceived and experimented with in the late 1940s and have long been used to recognize patterns. However, the last few years have seen a number of dramatic breakthroughs that have resulted in significant advances in performance, especially when multiple layers of neurons are employed—a technology that has come to be called "deep learning." Deep learning systems already power the speech recognition capability in Apple's Siri and are poised to accelerate progress in a broad range of applications that rely on pattern analysis and recognition. A deep learning neural network designed in 2011 by scientists at the University of Lugano in Switzerland, for example, was able to correctly identify more than 99 percent of the images in a large database of traffic signs—a level of accuracy that exceeded that of human experts who competed against the system. Researchers at Facebook have likewise developed an experimental system—consisting of nine levels of artificial neurons—that can correctly determine whether two photographs are of the same person 97.25 percent of the time, even if lighting conditions and orientation of the faces vary. That compares with 97.53 percent accuracy for human observers.[9]

Geoffrey Hinton of the University of Toronto, one of the leading researchers in the field, notes that deep learning technology "scales beautifully. Basically you just need to keep making it bigger and faster, and it will get better."[10] In other words, even without accounting for likely future improvements in their design, machine learning systems powered by deep learning networks are virtually certain to see continued dramatic progress simply as a result of Moore's Law.

Big data and the smart algorithms that accompany it are having an immediate impact on workplaces and careers as employers, particularly large corporations, increasingly track a myriad of metrics and statistics regarding the work and social interactions of their employees. Companies are relying ever more on so-called people analytics as a way to hire, fire, evaluate, and promote workers. The amount of data being collected on individuals and the work they engage in is staggering. Some companies capture every keystroke typed by every employee. Emails, phone records, web searches, database queries and accesses to files, entry and exit from facilities, and untold numbers of other types of data may also be collected—with or without the knowledge of workers.[11] While the initial purpose of all this data collection and analysis is typically more effective management and assessment of employee performance, it could eventually be put to other uses—including the development of software to automate much of the work being performed.

The big data revolution is likely to have two especially important implications for knowledge-based occupations. First, the data captured may, in many cases, lead to direct automation of specific tasks and jobs. Just as a person might study the historical record and then practice completing specific tasks in order to learn a new job, smart algorithms will often succeed using essentially the same approach. Consider, for example, that in November 2013 Google applied for a patent on a system designed to automatically generate personalized email and social media responses.[12] The system works by first analyzing a person's past emails and social media interactions. Based on what it learned, it would then automatically write responses to future emails, Tweets, or blog posts, and it would do so employing the person's usual writing style and tone. It's easy to imagine such a system eventually being used to automate a great deal of routine communication.

Google's automated cars, which it first demonstrated in 2011, likewise provide important insight into the path that data-driven automation is likely to follow. Google didn't set out to replicate the way a person drives—in fact, that would have been beyond the current capabilities of artificial intelligence. Rather, it simplified the challenge by designing a powerful data processing system and then putting it on wheels. Google's cars navigate by relying on precision location awareness via GPS together with vast amounts of extremely detailed mapping data. The cars also, of course, have radars, laser range finders, and other systems that provide a continuous stream of real-time information and allow the car to adapt to new situations, such as a pedestrian stepping off the curb. Driving may not be a white-collar profession, but the general strategy used by Google can be extended into a great many other areas: First, employ massive amounts of historical data in order to create a general "map" that will allow algorithms to navigate their way through routine tasks. Next, incorporate self-learning systems that can adapt to variations or unpredictable situations. The result is likely to be smart software that can perform many knowledge-based jobs with a high degree of reliability.

The second, and probably more significant, impact on knowledge jobs will occur as a result of the way big data changes organizations and the methods by which they are managed. Big data and predictive algorithms have the potential to transform the nature and number of knowledge-based jobs in organizations and industries across the board. The predictions that can be extracted from data will increasingly be used to substitute for human qualities such as experience and judgment. As top managers increasingly employ data-driven decision making powered by automated tools, there will be an ever-shrinking need for an extensive human analytic and management infrastructure. Whereas today there is a team of knowledge workers who collect information and present analysis to multiple levels of management, eventually there may be a single manager and a powerful algorithm. Organizations are likely to flatten. Layers of middle management will evaporate, and many of the jobs now performed by both clerical workers and skilled analysts will simply disappear.

WorkFusion, a start-up company based in the New York City area, offers an especially vivid example of the dramatic impact that white-collar automation is likely to have on organizations. The company offers large corporations an intelligent software platform that almost completely manages the execution of projects that were once highly labor-intensive through a combination of crowd sourcing and automation.

The WorkFusion software initially analyzes the project to determine which tasks can be directly automated, which can be crowd sourced, and which must be performed by in-house professionals. It can then automatically post job listings to websites like Elance or Craigslist and manage the recruitment and selection of qualified freelance workers. Once the workers are on board, the software allocates tasks and evaluates performance. It does this in part by asking freelancers to answer questions to which it already knows the answer as an ongoing test of the workers' accuracy. It tracks productivity metrics like typing speed, and automatically matches tasks with the capabilities of individuals. If a particular person is unable to complete a given assignment, the system will automatically escalate that task to someone with the necessary skills.

While the software almost completely automates management of the project and dramatically reduces the need for in-house employees, the approach does, of course, create new opportunities for freelance workers. The story doesn't end there, however. As the workers complete their assigned tasks, WorkFusion's machine learning algorithms continuously look for opportunities to further automate the process. In other words, even as the freelancers work under the direction of the system, they are simultaneously generating the training data that will gradually lead to their replacement with full automation.

One of the company's initial projects involved retrieving the information necessary to update a collection of about 40,000 records. Previously, the corporate client had performed this process annually using an in-house staff at a cost of nearly $4 per record. After switching to the WorkFusion platform, the client was able to update the records monthly at a cost of just 20 cents each. WorkFusion has found that, as the system's machine learning algorithms incrementally automate the process further, costs typically drop by about 50 percent after one year and still another 25 percent after a second year of operation.[13]

Cognitive Computing and IBM Watson

In the fall of 2004, IBM executive Charles Lickel had dinner with a small team of researchers at a steakhouse near Poughkeepsie, New York. Members of the group were taken aback when, at precisely seven o'clock, people suddenly began standing up from their tables and crowding around a television in the bar area. It turned out that Ken Jennings, who had already won more than fifty straight matches on the TV game show *Jeopardy!*, was once again attempting to extend his historic winning streak. Lickel noticed that the restaurant's patrons were so engaged that they abandoned their dinners, returning to finish their steaks only after the match concluded.[14]

That incident, at least according to many recollections, marked the genesis of the idea to build a computer capable of playing—and beating the very best human champions at—*Jeopardy!*[*] IBM had a long history of investing in high-profile projects called "grand challenges"

---

* Stephen Baker's 2011 book, *Final Jeopardy: Man vs. Machine and the Quest to Know Everything*, offers a detailed account of the fascinating story that ultimately led to IBM's Watson.

that have showcased the company's technology while delivering the kind of organic marketing buzz that just can't be purchased at any price. In a previous grand challenge, more than seven years earlier, IBM's Deep Blue computer had defeated world chess champion Garry Kasparov in a six-game match—an event that forever anchored the IBM brand to the historic moment when a machine first achieved dominance in the game of chess. IBM executives wanted a new grand challenge that would captivate the public and position the company as a clear technology leader—and, in particular, combat any perception that the information technology innovation baton had passed from Big Blue to Google or to start-up companies emerging out of Silicon Valley.

As the idea for a *Jeopardy!*-based grand challenge that would culminate in a televised match between the best human competitors and an IBM computer began to gain traction with the company's top managers, the computer scientists who would have to actually build such a system initially pushed back aggressively. A *Jeopardy!* computer would require capabilities far beyond anything that had been demonstrated previously. Many researchers feared that the company risked failure or, even worse, embarrassment on national television.

Indeed, there was little reason to believe that Deep Blue's triumph at chess would be extensible to *Jeopardy!* Chess is a game with precise rules that operate within a strictly limited domain; it is almost ideally suited to a computational approach. To a significant extent, IBM succeeded simply by throwing powerful, customized hardware at the problem. Deep Blue was a refrigerator-sized system packed with processors that were designed specifically for playing chess. "Brute force" algorithms leveraged all that computing power by considering every conceivable move given the current state of the game. Then for each of those possibilities, the software looked many moves ahead, weighing potential actions by both players and iterating through countless permutations—a laborious process that ultimately nearly always produced the optimal course of action. Deep Blue was fundamentally an exercise in pure mathematical calculation; all the information the computer needed to play the game was provided in a machine-friendly format it could process directly. There was no requirement for the machine to engage with its environment like a human chess player.

Jeopardy! presented a dramatically different scenario. Unlike chess, it is essentially open-ended. Nearly any subject that would be accessible to an educated person—science, history, film, literature, geography, and popular culture, to name just a few—is fair game. A computer would also face an entire range of daunting technical challenges. Foremost among these was the need to comprehend natural language: the computer would have to receive information and provide its responses in the same format as its human competitors. The hurdle for succeeding at *Jeopardy!* is especially high because the show has to be not just a fair contest but also an engaging form of entertainment for its millions of television viewers. The show's writers often intentionally weave humor, irony, and subtle plays on words into the clues—in other words, the kind of inputs that seem almost purposely designed to elicit ridiculous responses from a computer.

As an IBM document describing the Watson technology points out: "We have noses that run, and feet that smell. How can a slim chance and a fat chance be the same, but a wise man and a wise guy are opposites? How can a house burn up as it burns down? Why do we fill in a form by filling it out?"[15] A *Jeopardy!* computer would have to successfully navigate routine language ambiguities of that type while also exhibiting a level of general understanding far beyond what you'd typically find in computer algorithms designed to delve into mountains of text and retrieve relevant answers. As an example, consider the clue "Sink it & you've scratched." That clue was presented in a show televised in July 2000 and appeared on the top row of the game board—meaning that it was considered to be very easy. Try searching for that phrase using Google, and you'll get page after page of links to web pages about removing scratches from stainless-steel kitchen sinks. (That's assuming you exclude the exact match on a website about past *Jeopardy!* matches.) The correct response—"What is the cue ball?"—completely eludes Google's keyword-based search algorithm.*

All these challenges were well understood by David Ferrucci, the artificial intelligence expert who eventually assumed leadership of the team that built Watson. Ferrucci had previously managed a small group of IBM researchers focused on building a system that could answer questions provided in natural language format. The team entered their system, which they named "Piquant," in a contest run by the National Bureau of Standards and Technology—the same government agency that sponsored the machine language contest in which Google prevailed. In the contest, the competing systems had to churn through a defined set of about a million documents and come up with the answers to questions, and they were subject to no time limit at all. In some cases, the algorithms would grind away for several minutes before returning an answer.[16] This was a dramatically easier challenge than playing *Jeopardy!*, where the clues could draw on a seemingly limitless body of knowledge and where the machine would have to generate consistently correct responses within a few seconds in order to have any chance against top human players.

Piquant (as well as its competitors) was not only slow; it was inaccurate. The system was able to answer questions correctly only about 35 percent of the time—not an appreciably better success rate than you could get by simply typing the question into Google's search engine.[17] When Ferrucci's team tried to build a prototype *Jeopardy!*-playing system based on the Piquant project, the results were uniformly dismal. The idea that Piquant might someday take on a top *Jeopardy!* competitor like Ken Jennings seemed laughable. Ferrucci

---

* In *Jeopardy!* the clues are considered to be answers and the response must be phrased as a question for which the provided answer would be correct.

recognized that he would have to start from scratch—and that the project would be a major undertaking spanning as much as half a decade. He received the green light from IBM management in 2007 and set out to build, in his words, "the most sophisticated intelligence architecture the world has ever seen."[18] To do this, he drew on resources from throughout the company and put together a team consisting of artificial intelligence experts from within IBM as well as at top universities, including MIT and Carnegie Mellon.[19]

Ferrucci's team, which eventually grew to include about twenty researchers, began by building a massive collection of reference information that would form the basis for Watson's responses. This amounted to about 200 million pages of information, including dictionaries and reference books, works of literature, newspaper archives, web pages, and nearly the entire content of Wikipedia. Next they collected historical data for the *Jeopardy!* quiz show. Over 180,000 clues from previously televised matches became fodder for Watson's machine learning algorithms, while performance metrics from the best human competitors were used to refine the computer's betting strategy.[20] Watson's development required thousands of separate algorithms, each geared toward a specific task—such as searching within text; comparing dates, times, and locations; analyzing the grammar in clues; and translating raw information into properly formatted candidate responses.

Watson begins by pulling apart the clue, analyzing the words, and attempting to understand what exactly it should look for. This seemingly simple step can, in itself, be a tremendous challenge for a computer. Consider, for example, a clue that appeared in a category entitled "Lincoln Blogs" and was used in training Watson: "Secretary Chase just submitted this to me for the third time; guess what, pal. This time I'm accepting it." In order to have any chance at responding correctly, the machine would first need to understand that the initial instance of the word "this" acts as a placeholder for the answer it should seek.[21]

Once it has a basic understanding of the clue, Watson simultaneously launches hundreds of algorithms, each of which takes a different approach as it attempts to extract a possible answer from the massive corpus of reference material stored in the computer's memory. In the example above, Watson would know from the category that "Lincoln" is important, but the word "blogs" would likely be a distraction: unlike a human, the machine wouldn't comprehend that the show's writers were imagining Abraham Lincoln as a blogger.

As the competing search algorithms reel in hundreds of possible answers, Watson begins to rank and compare them. One technique used by the machine is to plug the potential answer into the original clue so that it forms a statement, and then go back out to the reference material and look for corroborating text. So if one of the search algorithms manages to come up with the correct response "resignation," Watson might then search its dataset for a statement something like "Secretary Chase just submitted resignation to Lincoln for the third time." It would find plenty of close matches, and the computer's confidence in that particular answer would rise. In ranking its candidate responses, Watson also relies on reams of historical data; it knows precisely which algorithms have the best track records for various types of questions, and it listens far more attentively to the top performers. Watson's ability to rank correctly worded natural language answers and then determine whether or not it has sufficient confidence to press the *Jeopardy!* buzzer is one of the system's defining characteristics, and a quality that places it on the frontier of artificial intelligence. IBM's machine "knows what it knows"—something that comes easily to humans but eludes nearly all computers when they delve into masses of unstructured information intended for people rather than machines.

Watson prevailed over *Jeopardy!* champions Ken Jennings and Brad Rutter in two matches televised in February 2011, giving IBM the massive publicity surge it hoped for. Well before the media frenzy surrounding that remarkable accomplishment began to fade, a far more consequential story began to unfold: IBM launched its campaign to leverage Watson's capabilities in the real world. One of the most promising areas is in medicine. Repurposed as a diagnostic tool, Watson offers the ability to extract precise answers from a staggering amount of medical information that might include textbooks, scientific journals, clinical studies, and even physicians' and nurses' notes for individual patients. No single doctor could possibly approach Watson's ability to delve into vast collections of data and discover relationships that might not be obvious—especially if the information is drawn from sources that cross boundaries between medical specialties.* By 2013, Watson was helping to diagnose problems and refine patient treatment plans at major medical facilities, including the Cleveland Clinic and the University of Texas's MD Anderson Cancer Center.

As a part of their effort to turn Watson into a practical tool, IBM researchers confronted one of the primary tenets of the big data revolution: the idea that prediction based on correlation is sufficient, and that a deep understanding of causation is usually both unachievable and unnecessary. A new feature they named "WatsonPaths" goes beyond simply providing an answer and lets researchers see the specific sources Watson consulted, the logic it used in its evaluation, and the inferences it made on its way to generating an answer. In other words, Watson is gradually progressing toward offering more insight into *why* something is true. WatsonPaths is also being used as a tool to help train medical students in diagnostic techniques. Less

* According to Stephen Baker's 2011 book *Final Jeopardy*, the Watson project leader, David Ferrucci, struggled with intense pain in one of his teeth for months. After multiple visits to dentists and what ultimately proved to be a completely unnecessary root canal, Ferrucci was finally—largely by happenstance—referred to a doctor in a medical specialty unrelated to dentistry, and the problem was solved. The specific condition was also described in a relatively obscure medical journal article. It was not lost on Ferrucci that a machine like Watson might have produced the correct diagnosis almost instantly.

than three years after a team of humans succeeded in building and training Watson, the tables have—at least to a limited extent—been turned, and people are now learning from the way the system reasons when presented with a complex problem.[22]

Other obvious applications for the Watson system are in areas like customer service and technical support. In 2013, IBM announced that it would work with Fluid, Inc., a major provider of online shopping services and consulting. The project aims to let online shopping sites replicate the kind of personalized, natural language assistance you would get from a knowledgeable sales clerk in a retail store. If you're going camping and need a tent, you'd be able to say something like "I am taking my family camping in upstate NY in October and I need a tent. What should I consider?" You'd then get specific tent recommendations, as well as pointers to other items that you might not have considered.[23] As I suggested in Chapter 1, it is only a matter of time before capability of that type becomes available via smart phones and shoppers are able to access conversational, natural language assistance while in brick and mortar stores.

MD Buyline, Inc., a company that specializes in providing information and research about the latest health care technology to hospitals, likewise plans to use Watson to answer the far more technical questions that come up when hospitals need to purchase new equipment. The system would draw on product specifications, prices, and clinical studies and research to make specific and instant recommendations to doctors and procurement managers.[24] Watson is also looking for a role in the financial industry, where the system may be poised to provide personalized financial advice by delving into a wealth of information about specific customers as well as general market and economic conditions. The deployment of Watson in customer service call centers is perhaps the area with the most disruptive near-term potential, and it is likely no coincidence that within a year of Watson's triumph on *Jeopardy!*, IBM was already working with Citigroup to explore applications for the system in the company's massive retail banking operation.[25]

IBM's new technology is still in its infancy. Watson—as well as the competing systems that are certain to eventually appear—have the potential to revolutionize the way questions are asked and answered, as well as the way information analysis is approached, both internal to organizations and in engagements with customers. There is no escaping the reality, however, that a great deal of the analysis performed by systems of this type would otherwise have been done by human knowledge workers.

Building Blocks in the Cloud

In November 2013, IBM announced that its Watson system would move from the specialized computers that hosted the system for the *Jeopardy!* matches to the cloud. In other words, Watson would now reside in massive collections of servers connected to the Internet. Developers would be able to link directly to the system and incorporate IBM's revolutionary cognitive computing technology into custom software applications and mobile apps. This latest version of Watson was also more than twice as fast as its *Jeopardy!*-playing predecessor. IBM envisions the rapid emergence of an entire ecosystem of smart, natural language applications—all carrying the "Powered by Watson" label.[26]

The migration of leading-edge artificial intelligence capability into the cloud is almost certain to be a powerful driver of white-collar automation. Cloud computing has become the focus of intense competition among major information technology companies, including Amazon, Google, and Microsoft. Google, for example, offers developers a cloud-based machine learning application as well as a large-scale compute engine that lets developers solve huge, computationally intensive problems by running programs on massive supercomputer-like networks of servers. Amazon is the industry leader in providing cloud computing services. Cycle Computing, a small company that specializes in large-scale computing, was able to solve a complex problem that would have taken over 260 years on a single computer in just 18 hours by utilizing tens of thousands of the computers that power Amazon's cloud service. The company estimates that prior to the advent of cloud computing, it would have cost as much as $68 million to build a supercomputer capable of taking on the problem. In contrast, it's possible to rent 10,000 servers in the Amazon cloud for about $90 per hour.[27]

Just as the field of robotics is poised for explosive growth as the hardware and software components used in designing the machines become cheaper and more capable, a similar phenomenon is unfolding for the technology that powers the automation of knowledge work. When technologies like Watson, deep learning neural networks, or narrative-writing engines are hosted in the cloud, they effectively become building blocks that can be leveraged in countless new ways. Just as hackers quickly figured out that Microsoft's Kinect could be used as an inexpensive way to give robots three-dimensional machine vision, developers will likewise find unforeseen—and perhaps revolutionary—applications for cloud-based software building blocks. Each of these building blocks is in effect a "black box"—meaning that the component can be used by programmers who have no detailed understanding of how it works. The ultimate result is sure to be that groundbreaking AI technologies created by teams of specialists will rapidly become ubiquitous and accessible even to amateur coders.

While innovations in robotics produce tangible machines that are often easily associated with particular jobs (a hamburger-making robot or a precision assembly robot, for example), progress in software automation will likely be far less visible to the public; it will often take place deep within corporate walls, and it will have more holistic impacts on organizations and the people they employ. White-collar automation will very often be the story of information technology consultants descending on large organizations and building completely custom systems that have the potential to revolutionize the way the business operates, while at the same time eliminating the need for potentially hundreds or even thousands of skilled workers. Indeed, one of IBM's stated motivations for creating the Watson technology was to offer its consulting division—which, together with software sales, now accounts for the vast majority of the company's revenues—a competitive advantage. At the same time, entrepreneurs are already finding ways to use the same cloud-based building blocks to create affordable automation products geared toward small or medium-sized businesses.

Cloud computing has already had a significant impact on information technology jobs. During the 1990's tech boom, huge numbers of well-paying jobs were created as businesses and organizations of all sizes needed IT professionals to administer and install personal computers, networks, and software. By the first decade of the twenty-first century, however, the trend began to shift as companies were increasingly outsourcing many of their information technology functions to huge, centralized computing hubs.

The massive facilities that host cloud computing services benefit from enormous economies of scale, and the administrative functions that once kept armies of skilled IT workers busy are now highly automated. Facebook, for example, employs a smart software application called "Cyborg" that continuously monitors tens of thousands of servers, detects problems, and in many cases can perform repairs completely autonomously. A Facebook executive noted in November 2013 that the Cyborg system routinely solves thousands of problems that would otherwise have to be addressed manually, and that the technology allows a single technician to manage as many as 20,000 computers.[28]

Cloud computing data centers are often built in relatively rural areas where land and, especially, electric power are plentiful and cheap. States and local governments compete intensively for the facilities, offering companies like Google, Facebook, and Apple generous tax breaks and other financial incentives. Their primary objective, of course, is to create lots of jobs for local residents—but such hopes are rarely realized. In 2011, the *Washington Post*'s Michael Rosenwald reported that a colossal, billion-dollar data center built by Apple, Inc., in the town of Maiden, North Carolina, had created only fifty full-time positions. Disappointed residents couldn't "comprehend how expensive facilities stretching across hundreds of acres can create so few jobs."[29] The explanation, of course, is that algorithms like Cyborg are doing the heavy lifting.

The impact on employment extends beyond the data centers themselves to the companies that leverage cloud computing services. In 2012, Roman Stanek, the CEO of Good Data, a San Francisco company that uses Amazon's cloud services to perform data analysis for about 6,000 clients, noted that "[b]efore, each [client] company needed at least five people to do this work. That is 30,000 people. I do it with 180. I don't know what all those other people will do now, but this isn't work they can do anymore. It's a winner-takes-all consolidation."[30]

The evaporation of thousands of skilled information technology jobs is likely a precursor for a much more wide-ranging impact on knowledge-based employment. As Netscape co-founder and venture capitalist Marc Andreessen famously said, "Software is eating the world." More often than not, that software will be hosted in the cloud. From that vantage point it will eventually be poised to invade virtually every workplace and swallow up nearly any white-collar job that involves sitting in front of a computer manipulating information.

Algorithms on the Frontier

If there is one myth regarding computer technology that ought to be swept into the dustbin it is the pervasive believe that computers can do only what they are specifically programmed to do. As we've seen, machine learning algorithms routinely churn through data, revealing statistical relationships and, in essence, writing their own programs on the basis of what they discover. In some cases, however, computers are pushing even further and beginning to encroach into areas that nearly everyone assumes are the exclusive province of the human mind: machines are starting to demonstrate curiosity and creativity.

In 2009, Hod Lipson, the director of the Creative Machines Lab at Cornell University, and PhD student Michael Schmidt built a system that has proved capable of independently discovering fundamental natural laws. Lipson and Schmidt started by setting up a double pendulum—a contraption that consists of one pendulum attached to, and dangling below, another. When both pendulums are swinging, the motion is extremely complex and seemingly chaotic. Next they used sensors and cameras to capture the pendulum's motion and produce a stream of data. Finally, they gave their software the ability to control the starting position of the pendulum; in other words, they created an artificial scientist with the ability to conduct its own experiments.

They turned their software loose to repeatedly release the pendulum and then sift through the resulting motion data and try to figure out the mathematical equations that describe the pendulum's behavior. The algorithm had complete control over the experiment; for each repetition, it decided how to position the pendulum for release, and it did not do this randomly—it performed an analysis and then chose the specific starting point that would likely provide the most insight into the laws underlying the pendulum's motion. Lipson notes that the system "is not a passive algorithm that sits back, watching. It *asks questions*. That's *curiosity*."[31] The program, which they later named "Eureqa," took only a few hours to come up with a number of physical laws describing the movement of the pendulum—including Newton's Second Law—and it was able to do this without being given any prior information or programming about physics or the laws of motion.

Eureqa uses genetic programming, a technique inspired by biological evolution. The algorithm begins by randomly combining various mathematical building blocks into equations and then testing to see how well the equations fit the data.* Equations that fail the test are discarded, while those that show promise are retained and recombined in new ways so that the system ultimately converges on an accurate mathematical model.[32] The process of finding an equation that describes the behavior of a natural system is by no means a trivial exercise. As Lipson says, "[P]reviously, coming up with a predictive model could take a [scientist's] whole career."[33] Schmidt adds that "[p]hysicists like Newton and Kepler could have used a computer running this algorithm to figure out the laws that explain a falling apple or the motion of the planets with just a few hours of computation."[34]

When Schmidt and Lipson published a paper describing their algorithm, they were deluged with requests for access to the software from other scientists, and they decided to make Eureqa available over the Internet in late 2009. The program has since produced a number of useful results in a range of scientific fields, including a simplified equation describing the biochemistry of bacteria that scientists are still struggling to understand.[35] In 2011, Schmidt founded Nutonian, Inc., a Boston-area start-up company focused on commercializing Eureqa as a big data analysis tool for both business and academic applications. One result is that Eureqa—like IBM's Watson—is now hosted in the cloud and is available as an application building block to other software developers.

Most of us quite naturally tend to associate the concept of creativity exclusively with the human brain, but it's worth remembering that the

* This is significantly more advanced than the commonly used statistical technique known as "regression." With regression (either linear or nonlinear), the form of the equation is set in advance, and the equation's parameters are optimized so as to fit the data. The Eureqa program, in contrast, is able to independently determine equations of any form using a variety of mathematical components including arithmetic operators, trigonometric and logarithmic functions, constants, etc.

brain itself—by far the most sophisticated invention in existence—is the product of evolution. Given this, perhaps it should come as no surprise that attempts to build creative machines very often incorporate genetic programming techniques. Genetic programming essentially allows computer algorithms to design themselves through a process of Darwinian natural selection. Computer code is initially generated randomly and then repeatedly shuffled using techniques that emulate sexual reproduction. Every so often, a random mutation is thrown in to help drive the process in entirely new directions. As new algorithms evolve, they are subjected to a fitness test that leads to either their survival, or—far more often—their demise. Computer scientist and consulting Stanford professor John Koza is one of the leading researchers in the field and has done extensive work using genetic algorithms as "automated invention machines."* Koza has isolated at least seventy-six cases where genetic algorithms have produced designs that are competitive with the work of human engineers and scientists in a variety of fields, including electric circuit design, mechanical systems, optics, software repair, and civil engineering. In most of these cases, the algorithms have replicated existing designs, but there are at least two instances where genetic programs have created new, patentable inventions.[36] Koza argues that genetic algorithms may have an important advantage over human designers because they are not constrained by preconceptions; in other words, they may be more likely to result in an "outside-the-box" approach to the problem.[37]

Lipson's suggestion that Eureqa exhibits curiosity and Koza's argument about computers acting without preconceptions suggest that creativity may be something that is within reach of a computer's

* In addition to his work in genetic programming, Koza is the inventor of the scratch-off lottery ticket and the originator of the "constitutional workaround" idea to elect US presidents by popular vote by having the states agree to award electoral-college votes based on the country's overall popular-vote outcome.

capabilities. The ultimate test of such an idea might be to see if a computer could create something that humans would accept as a work of art. Genuine artistic creativity—perhaps more so than any other intellectual endeavor—is something we associate exclusively with the human mind. As *Time*'s Lev Grossman says, "Creating a work of art is one of those activities we reserve for humans and humans only. It's an act of self-expression; you're not supposed to be able to do it if you don't have a self."[38] Embracing the possibility that a computer could be a legitimate artist would require a fundamental reevaluation of our assumptions about the nature of machines.

In the 2004 film *I, Robot,* the protagonist, played by Will Smith, asks a robot, "Can a robot write a symphony? Can a robot turn a canvas into a beautiful masterpiece?" The robot's reply "Can you?" is meant to suggest that, well, the vast majority of people can't do those things either. In the real world of 2015, however, Smith's question would elicit a more forceful answer: "Yes."

In July 2012, the London Symphony Orchestra performed a composition entitled *Transits—Into an Abyss.* One reviewer called it "artistic and delightful."[39] The event marked the first time that an elite orchestra had played music composed entirely by a machine. The composition was created by Iamus, a cluster of computers running a musically inclined artificial intelligence algorithm. Iamus, which is named after a character from Greek mythology who was said to understand the language of birds, was designed by researchers at the University of Malaga in Spain. The system begins with minimal information, such as the type of instruments that will play the music, and then, with no further human intervention, creates a highly complex composition—which can often evoke an emotional response in audiences—within minutes. Iamus has already produced millions of unique compositions in the modernist classical style, and is likely to be adapted to other musical genres in the future. Like Eureqa, Iamus has resulted in a start-up company to commercialize the technology. Melomics Media, Inc., has been set up to sell the music from an iTunes-like online store. The difference is that compositions created by Iamus are offered on a royalty-free basis, allowing purchasers to use the music in any way they wish.

Music is not the only art form being created by computers. Simon Colton, a professor of creative computing at the University of London, has built an artificial intelligence program called "The Painting Fool" that he hopes will someday be taken seriously as a painter (see Figure 4.1). "The goal of the project is not to produce software that can make photos look like they've been painted; Photoshop has done that for years," Colton says. "The goal is to see whether software can be accepted as creative in its own right."[40]

Colton has built a set of capabilities he calls "appreciative and imaginative behaviors" into the system. The Painting Fool software can identify emotions in photographs of people and then paint an abstract portrait that attempts to convey their emotional state. It can also generate imaginary objects using techniques based on genetic programming. Colton's software even has the ability to be self-critical. It does this by incorporating another software application called "Darci" that was built by researchers at Brigham Young

Figure 4.1. An Original Work of Art Created by Software

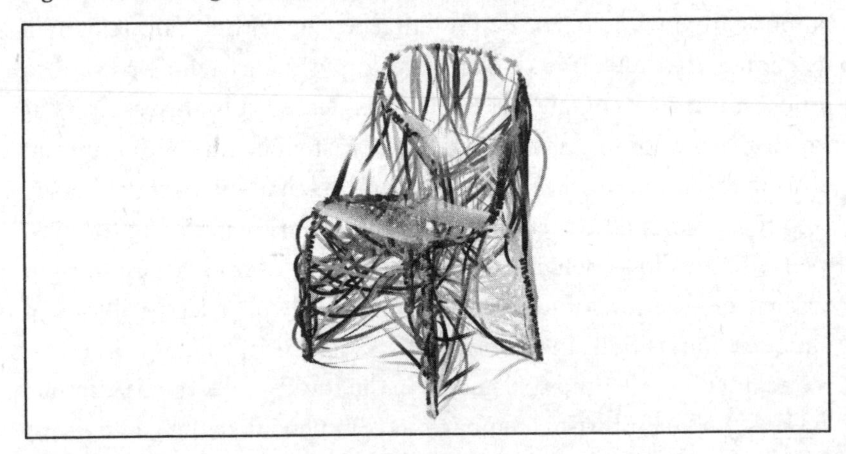

© ThePaintingFool.com

University. The Darci developers started with a database of paintings that had been labeled by humans with adjectives like "dark," "sad," or "inspiring." They then trained a neural network to make the associations and turned it loose to label new paintings. The Painting Fool is able to use feedback from Darci to decide whether or not it is achieving its objectives as it paints.[41]

My point here is not to suggest that large numbers of artists or musical composers will soon be out of a job. Rather, it is that the techniques used to build creative software—many of which, as we have seen, rely on genetic programming—can be repurposed in countless new ways. If computers can create musical compositions or design electronic components, then it seems likely that they will soon be able to formulate a new legal strategy or perhaps come up with a new way to approach a management problem. For the time being, the white-collar jobs at highest risk will continue to be those that are the most routine or formulaic—but the frontier is advancing quickly.

Nowhere is the rapid pace of that advance more evident than on Wall Street. Where once financial trading was highly dependent on direct communication between people, either in bustling trading pits or via telephone, it has now come to be largely dominated by machines communicating over fiber-optic links. By some estimates, automated trading algorithms are now responsible for at least half, and perhaps as much as 70 percent, of stock market transactions. These sophisticated robotic traders—many of which are powered by techniques on the frontier of artificial intelligence research—go far beyond simply executing routine trades. They attempt to profit by detecting and then snapping up shares in front of huge transactions initiated by mutual funds and pension managers. They seek to deceive other algorithms by inundating the system with decoy bids that are then withdrawn within tiny fractions of a second. Both Bloomberg and Dow News Service offer special machine-readable products designed to feed the algorithms' voracious appetites for financial news that they can—perhaps within milliseconds—turn into profitable trades. The news services also provide real-time metrics that let the machines see which items are attracting the most attention.[42] Twitter, Facebook, and the blogosphere are likewise all fodder for these competing algorithms. In a 2013 paper published in the scientific journal *Nature,* a group of physicists studied global financial markets and identified "an emerging ecology of competitive machines featuring 'crowds' of predatory algorithms," and suggested that robotic trading had progressed beyond the control—and even comprehension—of the humans who designed the systems.[43]

In the realm inhabited by these continuously battling algorithms, the action unfolds at a pace that would be incomprehensible to the fastest human trader. Indeed, speed—in some cases measured in millionths or even billionths of a second—is so critical to algorithmic trading success that Wall Street firms have collectively invested billions of dollars to build computing facilities and communications paths designed to produce tiny speed advantages. In 2009, for example, a company called Spread Networks spent as much as $200 million to lay down a new fiber-optic cable link stretching 825 miles in a straight line from Chicago to New York. The company operated in stealth mode so as not to alert the competition even as it blasted its way through the Allegheny Mountains. When the new fiber-optic path came online, it offered a speed advantage of perhaps three or four thousandths of a second compared with existing communications routes. That was enough to allow any algorithmic trading systems employing the new route to effectively dominate their competition. Wall Street firms, faced with algorithmic decimation, lined up to lease bandwidth—reportedly at a cost as much as ten times that of the original, slower cable. A similar cable stretching across the Atlantic between London and New York is currently in progress, and is expected to shave about five thousandths of a second off current execution times.[44]

The impact of all this automation is clear: even as the stock market continued on its upward trajectory in 2012 and 2013, large Wall

Street banks announced massive layoffs, often resulting in the elimination of tens of thousands of jobs. At the turn of the twenty-first century, Wall Street firms employed nearly 150,000 financial workers in New York City; by 2013, the number was barely more than 100,000—even as both the volume of transactions and the industry's profits soared.[45] Against the backdrop of that overall collapse in employment, Wall Street did create at least one very high-profile job: in late 2012, David Ferrucci, the computer scientist who led the effort to build Watson, left IBM for a new gig at a Wall Street hedge fund, where he'll be applying the latest advances in artificial intelligence to modeling the economy—and, presumably, trying to gain a competitive advantage for his firm's trading algorithms.[46]

Offshoring and High-Skill Jobs

While the trend toward increased automation of white-collar jobs is clear, the most dramatic onslaught—especially for truly skilled professions—still lies in the future. The same cannot necessarily be said for the practice of offshoring, where knowledge jobs are moved electronically to lower-wage countries. Highly educated and skilled professionals such as lawyers, radiologists, and especially computer programmers and information technology workers have already felt a significant impact. In India, for example, there are armies of call center workers and IT professionals, as well as tax preparers versed in the US tax code and attorneys specifically trained not in their own country's legal system but in American law, and standing ready to perform low-cost legal research for US firms engaged in domestic litigation. While the offshoring phenomenon may seem completely unrelated to the jobs lost to computers and algorithms, the precise opposite is true: offshoring is very often a precursor of automation, and the jobs it creates in low-wage nations may prove to be short-lived as technology advances. What's more, advances in artificial intelligence may make it even easier to offshore jobs that can't yet be fully automated.

Most economists view the practice of offshoring as just another example of global trade and argue that it invariably makes both parties to the transaction better off. Harvard professor N. Gregory Mankiw, for example, while serving as George W. Bush's chairman of the White House Council of Economic Advisers, said in 2004 that offshoring is "the latest manifestation of the gains from trade that economists have talked about at least since Adam Smith."[47] Abundant evidence argues to the contrary. Trade in tangible goods creates a great many peripheral jobs in areas like shipping, distribution, and retail. There are also natural forces that tend to mitigate the impact of globalization to some degree; for example, a company that chooses to move a factory to China incurs both shipping costs and a significant delay before completed products reach consumer markets. Electronic offshoring, in contrast, is almost completely frictionless and subject to none of these penalties. Jobs are moved to low-wage locations instantly and at minimal cost. If peripheral jobs are created, it is much more likely to be in the country where the workers reside.

I would argue that "free trade" is the wrong lens through which to view offshoring. Instead, it is much more akin to virtual immigration. Suppose, for example, that a huge customer service call center were to be built south of San Diego, just across the border from Mexico. Thousands of low-wage workers are issued "day worker" passes and are bused across the border to staff the call center every morning. At the end of the workday, the buses travel in the opposite direction. What is the difference between this situation (which would certainly be viewed as an immigration issue) and moving the jobs electronically to India or the Philippines? In both cases, workers are, in effect, "entering" the United States to offer services that are clearly directed at the domestic US economy. The biggest difference is that the Mexican day worker plan would probably be significantly better for the California economy. There might be jobs for bus drivers, and there would certainly be jobs for people to maintain the huge facility located on the US side of the border. Some of the workers might purchase lunch or even a cup of coffee while at work, thus injecting consumer demand into the local economy. The company that owned the California facility would pay property tax. When the jobs are offshored, and the workers enter the United States virtually, the domestic economy receives none of these benefits. I find it somewhat ironic that many conservatives in the United States are adamant about securing the border against immigrants who will likely take jobs that few Americans want, while at the same time expressing little concern that the virtual border is left completely open to higher-skill workers who take jobs that Americans definitely *do* want.

The argument put forth by economists like Mankiw, of course, measures in the aggregate and glosses over the highly disproportionate impact that offshoring has on the groups of people who either suffer or benefit from the practice. On the one hand, a relatively small but still significant group of people—potentially measured in the millions—may be subjected to a substantial downgrade in their income, quality of life, and future prospects. Many of these people may have made substantial investments in education and training. Some workers may lose their income entirely. Mankiw would likely argue that the aggregate benefit to consumers makes up for these losses. Unfortunately, although consumers may benefit from lower prices as a result of the offshoring, this savings may be spread across a population of tens or even hundreds of millions of people, perhaps resulting in a cost reduction that amounts to mere pennies and has a negligible effect on any one individual's well-being. And, needless to say, not all the gains will flow to consumers; a significant fraction will end up in the pockets of a few already-wealthy executives, investors, and business owners. This asymmetric impact is, perhaps not surprisingly, intuitively grasped by most average workers but seemingly lost on many economists.

One of the few economists to recognize offshoring's disruptive potential is the former vice chairman of the Federal Reserve's Board of Governors, Alan Blinder, who wrote a 2007 op-ed in the

Washington Post entitled "Free Trade's Great, but Offshoring Rattles Me."[48] Blinder has conducted a number of surveys aimed at assessing the future impact of offshoring and has estimated that 30–40 million US jobs—positions employing roughly a quarter of the workforce—are potentially offshorable. As he says, "We have so far barely seen the tip of the offshoring iceberg, the eventual dimensions of which may be staggering."[49]

Virtually any occupation that primarily involves manipulating information and is not in some way anchored locally—for example, with a requirement for face-to-face interaction with customers—is potentially at risk from offshoring in the relatively near future and then from full automation somewhat further out. Full automation is simply the logical next step. As technology advances, we can expect that more and more of the routine tasks now performed by offshore workers will eventually be handled entirely by machines. This has already occurred with respect to some call center workers who have been replaced by voice automation technology. As truly powerful natural language systems like IBM's Watson move into the customer service arena, huge numbers of offshore call center jobs are poised to be vaporized.

As this process unfolds, it seems likely that those companies—and nations—that have invested heavily in offshoring as a route to profitability and prosperity will have little choice but to move up the value chain. As more routine jobs are automated, higher-skill, professional jobs will be increasingly in the sights of the offshorers. One factor that is, I think, underappreciated is the extent to which advances in artificial intelligence as well as the big data revolution may act as a kind of catalyst, making a much broader range of high-skill jobs potentially offshorable. As we've seen, one of the tenets of the big data approach to management is that insights gleaned from algorithmic analysis can increasingly substitute for human judgment and experience. Even before advancing artificial intelligence applications reach the stage where full automation is possible, they will become powerful tools that encapsulate ever more of the analytic intelligence and institutional knowledge that give a business its competitive advantage. A smart young offshore worker wielding such tools might soon be competitive with far more experienced professionals in developed countries who command very high salaries.

When offshoring is viewed in combination with automation, the potential aggregate impact on employment is staggering. In 2013, researchers at the University of Oxford's Martin School conducted a detailed study of over seven hundred US job types and came to the conclusion that nearly 50 percent of jobs will ultimately be susceptible to full machine automation.[50] Alan Blinder and Alan Krueger of Princeton University conducted a similar analysis with respect to offshoring and found that about 25 percent of US jobs are at risk of eventually being moved to low-wage countries.[51] Let's hope there's significant overlap between those two estimates! Indeed, in all likelihood there is plenty of overlap when the estimates are viewed in terms of job titles or descriptions. The story is different along the time dimension, however. Offshoring will often arrive first; to a significant degree, it will accelerate the impact of automation even as it drags higher-skill jobs into the threat zone.

As powerful AI-based tools make it easier for offshore workers to compete with their higher-paid counterparts in developed countries, advancing technology is also likely to upend many of our most basic assumptions about which types of jobs are potentially offshorable. Nearly everyone believes, for example, that occupations that require physical manipulation of the environment will always be safe. Yet, military pilots located in the western United States routinely operate drone aircraft in Afghanistan. By the same token, it is easy to envision remote-controlled machinery being operated by offshore workers who provide the visual perception and dexterity that, for the time being, continues to elude autonomous robots. A need for face-to-face interaction is another factor that is assumed to anchor a job locally. However, telepresence robots are pushing the frontier in this area and have already been used to offshore English language instruction from Korean schools to the Philippines. In the not too distant future, advanced virtual reality environments will likewise make it even easier for workers to move seamlessly across national borders and engage directly with customers or clients.

As offshoring accelerates, college graduates in the United States and other advanced countries may face daunting competition based not just on wages but also on cognitive capability. The combined population of India and China amounts to roughly 2.6 billion people—or over eight times the population of the United States. The top 5 percent in terms of cognitive ability amounts to about 130 million people—or over 40 percent of the entire US population. In other words, the inescapable reality of the bell-curve distribution stipulates that there are far more very smart people in India and China than in the United States. That will not necessarily be a cause for concern, of course, as long as the domestic economies in those countries are capable of creating opportunities for all those smart workers. The evidence so far, however, suggests otherwise. India has built a major, nationally strategic industry specifically geared toward the electronic capture of American and European jobs. And China, even as the growth rate of its economy continues to be the envy of the world, struggles year after year to create sufficient white-collar jobs for its soaring population of new college graduates. In mid-2013, Chinese authorities acknowledged that only about half of the country's current crop of college graduates had been able to find jobs, while more than 20 percent of the previous year's graduates remained unemployed—and those figures are inflated when temporary and freelance work, as well as enrollment in graduate school and government-mandated "make work" positions, are regarded as full employment.[52]

Thus far, a lack of proficiency in English and other European languages has largely prevented skilled workers in China from competing aggressively in the offshoring industry. Once again, however, technology seems likely to eventually demolish this barrier.

Technologies like deep learning neural networks are poised to transport instantaneous machine voice translation from the realm of science fiction into the real world—and this could happen within the next few years. In June 2013, Hugo Barra, Google's top Android executive, indicated that he expects a workable "universal translator" that could be used either in person or over the phone to be available within several years. Barra also noted that Google already has "near perfect" real-time voice translation between English and Portuguese.[53] As more and more routine white-collar jobs fall to automation in countries throughout the world, it seems inevitable that competition will intensify to land one of the dwindling number of positions that remain beyond the reach of the machines. The very smartest people will have a significant advantage, and they won't hesitate to look beyond national borders. In the absence of barriers to virtual immigration, the employment prospects for nonelite college-educated workers in developed economies could turn out to be pretty grim.

Education and Collaboration with the Machines

As technology advances and more jobs become susceptible to automation, the conventional solution has always been to offer workers more education and training so that they can step into to new, higher-skill roles. As we saw in Chapter 1, millions of lower-skill jobs in areas like fast food and retail are at risk as robots and self-service technologies begin to encroach aggressively in these areas. We can be sure that more education and training will be the primary proffered solution for these workers. Yet, the message of this chapter has been that the ongoing race between technology and education may well be approaching the endgame: the machines are coming for the higher-skill jobs as well.

Among economists who are tuned in to this trend, a new flavor of conventional wisdom is arising: the jobs of the future will involve collaborating with the machines. Erik Brynjolfsson and Andrew McAfee of the Massachusetts Institute of Technology have been especially strong proponents of this idea, advising workers that they should learn to "race with the machines"—rather than against them.

While that may well be sage advice, it is nothing especially new. Learning to work with the prevailing technology has always been a good career strategy. We used to call it "learning computer skills." Nevertheless, we should be very skeptical that this latest iteration will prove to be an adequate solution as information technology continues on its relentless exponential path.

The poster child for the machine-human symbiosis idea has come to be the relatively obscure game of freestyle chess. More than a decade after IBM's Deep Blue computer defeated world chess champion Garry Kasparov, it is generally accepted that, in one-on-one contests between computers and humans, the machines now dominate absolutely. Freestyle chess, however, is a team sport. Groups of people, who are not necessarily world-class chess players individually, compete against each other and are allowed to freely consult with computer chess programs as they evaluate each move. As things stand in 2014, human teams with access to multiple chess algorithms are able to outmatch any single chess-playing computer.

There are a number of obvious problems with the idea that human-machine collaboration, rather than full automation, will come to dominate the workplaces of the future. The first is that the continued dominance of human-machine teams in freestyle chess is by no means assured. To me, the process that these teams use—evaluating and comparing the results from different chess algorithms before deciding on the best move—seems uncomfortably close to what IBM Watson does when it fires off hundreds of information-seeking algorithms and then succeeds in ranking the results. I don't think it is much of a stretch to suggest that a "meta" chess-playing computer with access to multiple algorithms may ultimately defeat the human teams—especially if speed is an important factor.

Secondly, even if the human-machine team approach does offer an incremental advantage going forward, there is an important question as to whether employers will be willing to make the investment necessary to leverage that advantage. In spite of the mottos and slogans that corporations direct at their employees, the reality is that most businesses are not prepared to pay a significant premium for "world-class" performance when it comes to the bulk of the more routine work required in their operations. If you have any doubts about this, I'd suggest trying to call your cable company. Businesses *will* make the investment in areas that are critical to their core competency—in other words, the activities that give the business a competitive advantage. Again, this scenario is nothing new. And, more importantly, it doesn't really involve any new people. The individuals that businesses are likely to hire and then couple with the best available technology are the same people who are largely immune to unemployment today. It is a small population of elite workers. Economist Tyler Cowen's 2013 book *Average Is Over* quotes one freestyle chess insider who says that the very best players are "genetic freaks."[54] That hardly makes the machine collaboration idea sound like a systemic solution for masses of people pushed out of routine jobs. And, as we have just seen, there is also the problem of offshoring. A great many of those 2.6 billion people in India and China are going to be pretty eager to grab one of those elite jobs.

There are also good reasons to expect that many machine collaboration jobs will be relatively short-lived. Recall the example of WorkFusion and how the company's machine learning algorithms incrementally automate the work performed by freelancers. The bottom line is that if you find yourself working with, or under the direction of, a smart software system, it's probably a pretty good bet that—whether you're aware of it or not—you are also training the software to ultimately replace you.

Yet another observation is that, in many cases, those workers who seek a machine collaboration job may well be in for a "be careful what you wish for" epiphany. As one example, consider the current trends in legal discovery. When corporations engage in litigation, it becomes necessary to sift through enormous numbers of internal documents and decide which ones are potentially relevant to the case at hand. The rules require these to be provided to the opposing side, and there can be substantial legal penalties for failing to produce anything that might be pertinent. One of the paradoxes of the paperless office is that the sheer number of such documents, especially in the form of emails, has grown dramatically since the days of typewriters and paper. To deal with this overwhelming volume, law firms are employing new techniques.

The first approach involves full automation. So-called e-Discovery software is based on powerful algorithms that can analyze millions of electronic documents and automatically tease out the relevant ones. These algorithms go far beyond simple key-word searches and often incorporate machine learning techniques that can isolate relevant concepts even when specific phrases are not present.[55] One direct result has been the evaporation of large numbers of jobs for lawyers and paralegals who once would have sorted laboriously through cardboard boxes full of paper documents.

There is also a second approach in common use: law firms may outsource this discovery work to specialists who hire legions of recent law school graduates. These graduates are typically victims of the bursting law school enrollment bubble. Unable to find employment as full-fledged lawyers—and often burdened with enormous student loans—they instead work as document reviewers. Each attorney sits in front of a monitor where a continuous stream of documents is displayed. Along with the document, there are two buttons: "Relevant" and "Not Relevant." The law school graduates scan the document on the screen and click the proper button. A new document then appears.[56] They may be expected to categorize up to eighty documents per hour.[57] For these young attorneys, there are no courtrooms, no opportunity to learn or to grow in their profession, and no opportunity for advancement. Instead, there are—hour after hour—the "Relevant" and "Not Relevant" buttons.*

One obvious question regarding these two competing approaches is whether the collaboration model is sustainable. Even at the relatively low wages (for attorneys) commanded by these workers, the automated approach seems far more cost-effective. As to the low quality of these jobs, you might assume that I've simply cherry-picked a rather dystopian example. After all, won't most jobs that involve collaboration with machines put people in control—so that workers supervise the machines and engage in rewarding work, rather than simply acting as gears and cogs in a mechanized process?

The problem with this rather wishful assumption is that the data does not support it. In his 2007 book *Super Crunchers,* Yale University professor Ian Ayres cites study after study showing that algorithmic approaches routinely outperform human experts. When people, rather than computers, are given overall control of the process, the results almost invariably suffer. Even when human experts are given access to the algorithmic results in advance, they *still* produce outcomes that are inferior to the machines acting autonomously. To the extent that people add value to the process, it is better to have them provide specific inputs to the system instead of giving them overall control. As Ayres says, "Evidence is mounting in favor of a different and much more demeaning, dehumanizing mechanism for combining expert and [algorithmic] expertise."[58]

---

* If you find this type of work appealing but lack the requisite legal training, be sure to check out Amazon's "Mechanical Turk" service, which offers many similar opportunities. "BinCam," for example, places cameras in your garbage bin, tracks everything you throw away, and then automatically posts the record to social media. The idea is, apparently, to shame yourself into not wasting food and not forgetting to recycle. As we've seen, visual recognition (of types of garbage, in this case) remains a daunting challenge for computers, so people are employed to perform this task. The very fact that this service is economically viable should give you some idea of the wage level for this kind of work.

My point here is that while human-machine collaboration jobs will certainly exist, they seem likely to be relatively few in number* and often short-lived. In a great many cases, they may also be unrewarding or even dehumanizing. Given this, it seems difficult to justify suggesting that we ought to make a major effort to specifically educate people in ways that will help them land one of these jobs—even if it were possible to pin down exactly what such training might entail. For the most part, this argument strikes me as a way to patch the tires on a very conventional idea (give workers still more vocational training) and keep it rolling for a bit longer. We are ultimately headed for a disruption that will demand a far more dramatic policy response.

SOME OF THE FIRST JOBS to fall to white-collar automation are sure to be the entry-level positions taken by new college graduates. As we saw in Chapter 2, there is already evidence to suggest that this process is well under way. Between 2003 and 2012, the median income of US college graduates with bachelor's degrees fell from nearly $52,000 to just over $46,000, measured in 2012 dollars. During the same period, total student loan debt tripled from about $300 billion to $900 billion.[59]

Underemployment among recent graduates is rampant, and nearly every college student seemingly knows someone whose degree has led to a career working at a coffee shop. In March 2013,

* In *Average Is Over,* Tyler Cowen estimates that perhaps 10–15 percent of the American workforce will be well equipped for machine collaboration jobs. I think that in the long run, even that estimate might be optimistic, especially when you consider the impact of offshoring. How many machine collaboration jobs will also be anchored locally? (One exception to my skepticism about machine collaboration jobs may be in health care. As discussed in Chapter 6, I think it might eventually be possible to create a new type of medical professional with far less training than a doctor who would work together with an AI-based diagnostic and treatment system. Health care is a special case, however, because doctors require an extraordinary amount of training and there is likely to be a significant shortage of physicians in the future.)

Canadian economists Paul Beaudry, David A. Green, and Benjamin M. Sand published an academic paper entitled "The Great Reversal in the Demand for Skill and Cognitive Tasks."[60] That title essentially says it all: the economists found that around the year 2000, overall demand for skilled labor in the United States peaked and then went into precipitous decline. The result is that new college graduates have increasingly been forced into relatively unskilled jobs—often displacing nongraduates in the process.

Even those graduates with degrees in scientific and technical fields have been significantly impacted. As we've seen, the information technology job market, in particular, has been transformed by the increased automation associated with the trend toward cloud computing as well as by offshoring. The widely held belief that a degree in engineering or computer science guarantees a job is largely a myth. An April 2013 analysis by the Economic Policy Institute found that at colleges in the United States, the number of new graduates with engineering and computer science degrees exceeds the number of graduates who actually find jobs in these fields by 50 percent. The study concludes that "the supply of graduates is substantially larger than the demand for them in industry."[61] It is becoming increasingly clear that a great many people will do all the right things in terms of pursuing an advanced education, but nonetheless fail to find a foothold in the economy of the future.

While some of the economists who focus their efforts on sifting through reams of historical data are finally beginning to discern the impact that advancing technology is having on higher-skill jobs, they are typically quite cautious about attempting to project that trend into the future. Researchers working in the field of artificial intelligence are often far less reticent. Noriko Arai, a mathematician with Japan's National Institute of Informatics, is leading a project to develop a system capable of passing the Tokyo University entrance examination. Arai believes that if a computer can demonstrate the combination of natural language aptitude and analytic skill necessary to gain entrance to Japan's highest-ranked university, then it will very likely also be able to eventually perform many of the jobs taken by college graduates. She foresees the possibility of massive job displacement within the next ten to twenty years. One of the primary motivations for her project is to try to quantify the potential impact of artificial intelligence on the job market. Arai worries that 10 to 20 percent of skilled workers replaced by automation would be a "catastrophe" and says she "can't begin to think what 50 percent would mean." She then adds that it would be "way beyond a catastrophe and such numbers can't be ruled out if AI performs well in the future."[62]

The higher-education industry itself has historically been one of the primary employment sectors for highly skilled workers. Especially for those who aspire to a doctoral degree, a typical career path has been to arrive on campus as a college freshman—and then never really leave. In the next chapter we'll look at how that industry, and a great many careers, may also be on the verge of a massive technological disruption.

Chapter 5

TRANSFORMING HIGHER EDUCATION

In March 2013, a small group of academics, consisting primarily of English professors and writing instructors, launched an online petition in response to news that essays on standardized tests were to be graded by machines. The petition, entitled "Professionals Against Machine Scoring of Student Essays in High Stakes Assessment,"[1] reflects the group's argument that algorithmic grading of written essays is, among other things, simplistic, inaccurate, arbitrary, and discriminatory, not to mention that it would be done "by a device that, in fact, cannot read." Within less than two months, the petition had been signed by nearly four thousand professional educators, as well as public intellectuals, including Noam Chomsky.

Using computers to grade tests is not new, of course; they've handled the trivial task of grading multiple-choice tests for years. In that context they are viewed as labor-saving devices. When the algorithms begin to encroach on an area believed to be highly dependent on human skill and judgment, however, many teachers see the technology as a threat. Machine essay grading draws on advanced artificial intelligence techniques; the basic strategy used to evaluate student essays is quite similar to the methodology behind Google's online language translation. Machine learning algorithms are first trained using a large number of writing samples that have already been graded by human instructors. The algorithms are then turned loose to score new student essays and are able to do so virtually instantaneously.

The "Professionals Against Machine Scoring" petition is certainly correct in its claim that the machines doing the grading "cannot read." As we've seen in other applications of big data and machine learning, however, that doesn't matter. Techniques based on the analysis of statistical correlations very often match or even outperform the best efforts of human experts. Indeed, a 2012 analysis by researchers at the University of Akron's College of Education compared machine grading with the scores awarded by human instructors and found that the technology "achieved virtually identical levels of accuracy, with the software in some cases proving to be more reliable." The study involved nine companies that offer machine grading solutions and over 16,000 pre-graded student essays from public school in six US states.[2]

Les Perelman, a former director of the Massachusetts Institute of Technology's writing program, is one of the most outspoken critics of machine grading, and one of the primary backers of the 2013 petition opposing the practice. Perelman has, in a number of cases, been able to construct completely nonsensical essays that have tricked the grading algorithms into awarding high scores. It seems to me, however, that if the skill required to put together rubbish designed to fool the software is roughly comparable to the skill needed to write a coherent essay, then this tends to undermine Perelman's argument that the system could be easily gamed. The real question is whether a student who lacks the ability to write effectively can put one over on the grading software, and the University of Akron study seems to suggest otherwise. Perelman does raise at least one valid concern, however: the prospect that students will be taught to write specifically to please algorithms that he suggests "disproportionately give students credit for length and loquacious wording."[3]

Algorithmic grading, despite the controversy that attaches to it, is virtually certain to become more prevalent as schools continue to seek ways to cut costs. In situations where a large number of essays need to be graded, the approach has obvious advantages. Aside from speed and lower cost, an algorithmic approach offers objectivity and consistency in cases where multiple human graders would otherwise be required. The technology also gives students instant feedback and is well suited to assignments that might not otherwise receive detailed scrutiny from an instructor. For example, many communications courses require or encourage students to maintain daily journals; an algorithm can evaluate each entry, and perhaps even suggest improvements, at the click of a button. It seems reasonable to assume that automated grading will, at least for the foreseeable future, be relegated to introductory courses teaching basic communication skills. English professors have little reason to fear that the algorithms are poised to invade upper-level creative writing seminars. However, their deployment in introductory courses might eventually displace the graduate teaching assistants who now perform these routine grading tasks.

The uproar over robotic essay grading represents only a small example of the backlash that is certain to arise as the full force of accelerating information technology finally falls upon the education sector. Thus far, colleges and universities have largely been immune to the substantial increases in productivity that have transformed other industries. The benefits of information technology have not yet scaled across the higher-education sector. This, at least in part, explains the extraordinary increase in the cost of college in recent decades.

There are strong indications that things are about to change. One of the most disruptive impacts is sure to come from online courses offered by elite institutions. In many cases, these courses attract huge enrollments, and they will, therefore, be an important driver of automated approaches to both teaching and grading. EdX, a consortium of elite universities founded to offer free online courses, announced in early 2013 that it will make its essay-grading software freely available to any educational institutions that want to use it.[4] In other words, algorithmic grading systems have become yet another example of an Internet-based software building block that will help accelerate the inevitable drive toward the increased automation of skilled human labor.

The Rise—and Stumble—of the MOOC

Free Internet-based courses like those offered by edX are part of the trend toward massive open online courses—or MOOCs—that exploded into the public consciousness in the late summer of 2011, when two computer scientists at Stanford University, Sebastian Thrun and Peter Norvig, announced that their introductory artificial intelligence class would be available to anyone at no cost over the Internet. Both of the course's instructors were celebrities in their field with strong ties to Google; Thrun had led the effort to develop the company's self-driving cars, while Norvig was the director of research and co-author of the leading AI textbook. Within days of the announcement, more than 10,000 people had signed up. When John Markoff of the *New York Times* wrote a front-page article[5] about the course that August, enrollment rocketed to more than 160,000 people from over 190 countries. The number of online students from Lithuania alone exceeded the entire undergraduate and graduate student enrollment at Stanford. Students as young as ten and as old as seventy signed up to learn the basics of AI directly from two of the field's preeminent researchers—an extraordinary opportunity previously available only to about 200 Stanford students.[6]

The ten-week course was divided into short segments lasting just a few minutes and modeled roughly on the enormously successful videos for middle and high school students created by the Khan

Academy. I completed several units of the class myself and found the format to be a powerful and engaging learning vehicle. The production employed no visual wizardry; instead, it consisted primarily of either Thrun or Norvig presenting topics while writing on a notepad. Each brief segment was followed by an interactive quiz—a technique that virtually guarantees that key concepts are assimilated as you proceed through the course. About 23,000 people completed the class, took the final exam, and received a statement of accomplishment from Stanford.

Within months, an entirely new industry materialized around the MOOC phenomenon. Sebastian Thrun rounded up venture capital and formed a new company named Udacity to offer free or low-cost online classes. Across the country and the globe, elite universities rushed to get in on the game. Two other Stanford professors, Andrew Ng and Daphne Koller, founded Coursera with a $22 million initial investment and built a partnership with Stanford, the University of Michigan, the University of Pennsylvania, and Princeton. Harvard and MIT quickly invested $60 million to form edX. Coursera responded by adding another dozen universities, including Johns Hopkins and the California Institute of Technology, and within eighteen months it was working with over a hundred institutions throughout the world.

By early 2013, the hype surrounding MOOCs was exploding as rapidly as course enrollments. The online classes were widely believed to be poised to usher in a new age in which elite education would be accessible to all at little or no cost. The poor throughout Africa and Asia would soon be attending Ivy League colleges via cheap tablets and smart phones. Columnist Thomas Friedman of the *New York Times* called MOOCs a "budding revolution in global online higher education" and suggested that the online courses had the potential to "unlock a billion more brains to solve the world's biggest problems."[7]

Reality struck in the form of two studies released by the University of Pennsylvania in the final months of 2013. One of the studies looked at a million people who had enrolled in classes offered by Coursera and found that MOOCs "have relatively few active users, that user 'engagement' falls off dramatically—especially after the first 1–2 weeks of a course—and that few users persist to the course end."[8] Only about half of the people who signed up for classes viewed even a single lecture. Course completion rates ranged from 2 to 14 percent, and averaged about 4 percent. MOOCs were also largely failing to attract the poor and undereducated students whom everyone thought stood to benefit the most; about 80 percent of the people who signed up for the classes already had a college degree.

Several months earlier, a high-profile partnership between Udacity and San Jose State University had likewise failed to measure up to expectations. The program, intended to offer disadvantaged students inexpensive online classes in remedial math, college algebra, and introductory statistics, was announced at a press conference by Sebastian Thrun and California governor Jerry Brown in January 2013 and touted as a possible solution to soaring tuition costs and overcrowding at state colleges. When the first groups of students completed the courses, which cost just $150 and offered online mentors to provide individual assistance, the results were dismal. Three-quarters of the students taking the algebra class—and nearly 90 percent of those coming directly from high school—failed the course. In general, the MOOC students did significantly worse than students enrolled in traditional classes at San Jose State. The university has since suspended the program at least temporarily.[9]

Udacity is now deemphasizing broad-based education and is instead focusing on more vocational classes designed to give workers specific technical skills. Companies like Google and Salesforce.com, for example, are underwriting courses that teach software developers how to work with their products. Udacity has also partnered with the Georgia Institute of Technology to offer the first MOOC-based master's degree in computer science. Tuition for the three-semester program will cost just $6,600—about 80 percent less than a traditional on-campus degree. The program's setup costs are being funded by AT&T, which plans to send many of its employees through the program. Initially, Georgia Tech will enroll about 375 students, but the goal is to expand the program so it can serve thousands.

As MOOCs continue to evolve and improve, the hope that they will drive a global revolution that will bring high-quality education to hundreds of millions of the world's poor may ultimately be realized. In the near term, however, it seems evident that these online courses are most likely to attract students who are already highly motivated to seek further education. In other words, MOOCs are poised to compete for the same people who might otherwise enroll in more traditional classes. Assuming that potential employers see MOOCs as offering a valuable credential, this could eventually unleash a dramatic disruption of the entire higher-education sector.

College Credit and Competency-Based Credentials

When Thrun and Norvig tallied up the results from their 2011 artificial intelligence class, they found that 248 participants had achieved perfect scores in the course; these students had never answered an examination question incorrectly. They also discovered that not a single Stanford student was among that elite group. In fact, the highest-scoring on-campus student was outperformed by at least 400 online participants. None of those stellar performers, however, received formal Stanford credit or even a traditional certificate of completion for their work.

Months earlier, when Stanford administrators had first learned of the course's soaring enrollment, they had repeatedly called the professors into meetings to negotiate the nature of any credential that might be awarded to the online participants. The concern was not just that the Stanford cachet would potentially be diluted across tens of thousands of people—none of whom was being charged the roughly $40,000 in annual tuition paid by regular on-campus students—but also that the identity of students in remote locations could not be verified. The administrators eventually agreed that a simple "statement of accomplishment" could be offered to students who completed the course over the Internet. Stanford officials were so concerned with this precise terminology that when a journalist used the word "certificate" in a column about the course, they immediately called to request a correction.

The Stanford officials' worries about verifying the identity of online students were not unfounded. Indeed, ensuring that credit is awarded to the same person who actually completes the course and takes the exams is one of the most significant challenges associated with offering college credit or official credentials for MOOCs. Without a robust identification process, a vibrant industry would soon spring up around the fraudulent completion of courses and exams. In fact, a number of websites have already appeared offering to take online courses for other people in return for a fee. In late 2012, journalists from the website *Inside Higher Ed* posed as students and requested information from some of these sites about completing an introductory online economics course offered by Penn State. They were quoted fees ranging from $775 to $900 and guaranteed at least a "B" in the course. And this was for a class at Penn State's traditional, degree-granting online branch, where verifying student identities ought to be far less of a challenge than it would be for an open class with enormous numbers of participants.[10] Total enrollment for the entire Penn State program is only about 6,000 graduate and undergraduate students—a tiny fraction of the number of people likely to sign up for a single popular MOOC.

Cheating has also been a significant problem with massive online classes. In 2012, dozens of complaints were filed about plagiarism in humanities courses offered through Coursera. These courses relied on peer grading, rather than algorithms, to assess student performance, so course administrators responding to complaints had to deal with both the possibility of rampant plagiarism and the likelihood that at least some of the accusations were erroneous. In one class on science fiction and fantasy writing, claims that student essays were being copied from Wikipedia or other previously published sources prompted Eric Rabkin, the University of Michigan English professor teaching the course, to send a letter to all 39,000 students warning them against appropriating others' work, but also pointing out that "an accusation of plagiarism is a deeply serious act and should be made only with concrete evidence behind it."[11] The remarkable thing about such incidents is that no academic credit is being offered for any of these classes. Apparently, some people will cheat "just because they can" or perhaps because they don't understand the rules. In any case, there can be little doubt that associating formal academic credit with such courses would dramatically increase the incentive to misbehave.

There are a number of possible technical solutions to the identification and cheating problems. A simple method is to pose challenge questions requesting personal data at the beginning of each session. If you're planning to cheat by hiring someone to take a class in your name, you might think twice before giving them your social security number. That type of strategy would be difficult to implement globally, however. A remote proctoring solution requires that a camera be active on the computer so administrators can monitor the student. In 2013, edX—the MOOC consortium founded by Harvard and MIT—began offering ID-verified certificates to students who pay an additional fee and take the class under the watchful eye of a webcam. Such certificates can be presented to potential employers but generally cannot be used for academic credit. Monitoring by proctors is expensive and obviously not extensible to tens of thousands of people taking a free course, but it seems likely that facial recognition algorithms of the type currently used to label photos on Facebook may eventually step into that role. Other algorithms may soon be able to identify students by analyzing the cadence of their keystrokes, or root out plagiarism by automatically comparing written assignments to vast datasets of existing works.[12]

An especially promising path toward attaching academic credit to MOOCs may be to offer competency-based credentials. With this approach, students earn credit not by attending a class but by passing separate assessment tests demonstrating competence in specific areas. Competency-based education (CBE) was pioneered at Western Governor's University (WGU), an online institution first proposed at a 1995 conference attended by the governors of nineteen western US states. WGU began operating in 1997 and by 2013 had over 40,000 students, many of whom are adults seeking to complete degree programs they started years earlier or to transition to new careers. The CBE approach received a major boost in September 2013, when the University of Wisconsin announced that it would introduce a competency-based program that will lead to degrees.

MOOCs and CBE may prove to be a natural fit because the combination essentially decouples the courses from the credential. Issues like student identification and cheating would have to be addressed only in the assessment tests. There may even be an opportunity for a venture-backed firm to step into the testing and credential issuing role while completely bypassing the messy and expensive business of offering classes. Self-motivated students would be free to use any available resources—including MOOCs, self-study, or more traditional classes—to achieve competency, and then could pass an assessment test administered by the firm for credit. Such tests might be quite rigorous, in effect creating a filter roughly comparable to the admissions processes at more selective colleges. If such a start-up company were able to build a solid reputation for granting credentials only to highly competent graduates, and if—perhaps most critically—it could build strong relationships with high-profile employers so that its graduates were sought after, it would have a clear potential to upend the higher-education industry.

An annual survey of top officials at nearly 3,000 US colleges and universities found that expectations regarding the future promise of MOOCs diminished significantly over the course of 2013. Nearly 40 percent of the survey's respondents said massive online courses were not a sustainable method of instruction; in the prior year's survey, only a quarter of college administrators had expressed that view. The *Chronicle of Higher Education* likewise offered a relatively grim progress report, noting that "MOOCs made no significant inroads in the past year in the existing credentialing system in higher education, calling into question whether they will be as disruptive to the status quo as some observers first thought."[13]

One of the paradoxes associated with MOOCs is that for all their practical problems as a mass education mechanism, they can be an enormously effective learning method for those students who have sufficient motivation and self-discipline. When Thrun and Norvig first began offering their artificial intelligence class online, they were surprised to see attendance at their Stanford lectures quickly begin to drop off, so that eventually only about 30 out of 200 on-campus students were showing up on a regular basis. Their students, it seemed, preferred to take the class online. They also found that the new MOOC format resulted in a significant boost in their on-campus students' average performance on exams, as compared with students who took the same class in prior years.

I think it would be very premature to declare the MOOC phenomenon down for the count. We may, rather, simply be seeing the early-stage stumbles that are typical of new technologies. It's worth remembering, for example, that Microsoft Windows did not mature into an industry-dominating force until Microsoft released version 3.0—at least five years after the product was first introduced. Indeed, pessimism regarding the future sustainability of MOOCs among college administrators is quite possibly tied in large measure to their fears about the economic impact these courses might potentially have on their institutions and the entire higher-education sector.

On the Brink of Disruption

If the MOOC disruption is yet to unfold, it will slam into an industry that brings in nearly half a trillion dollars in annual revenue and employs over three and a half million people.[14] In the years between 1985 and 2013, college costs soared by 538 percent, while the general consumer price index increased only 121 percent. Even medical costs lagged far behind higher education, increasing about 286 percent over the same period.[15] Much of that cost is being funded with student loans, which now amount to at least $1.2 trillion in the United States. About 70 percent of US college students borrow, and the average debt at graduation is just under $30,000.[16] Keep in mind that only about 60 percent of college students in bachelor's degree programs graduate within six years, leaving the remainder to pay off any accumulated debt without the benefit of a degree.[17]

Remarkably, the cost of actual instruction at colleges and universities has made a relatively minor contribution to these surging costs. In his 2013 book *College Unbound,* Jeffrey J. Selingo cites data gathered by the Delta Cost Project, a small research organization that produces highly regarded analysis for the higher-education industry. Between 2000 and 2010, large public research universities increased spending on student services by 19 percent, administration by 15 percent, and operations and maintenance by 20 percent. Lagging well behind was the cost of teaching, which rose just 10 percent.[18] At the University of California system, faculty employment actually fell by 2.3 percent between 2009 and 2011, even as student enrollment increased by 3.6 percent.[19] To keep instructional costs down, colleges are relying ever more heavily on part-time, or adjunct, faculty who are paid on a per-course basis—in some cases as little as $2,500 for a semester-long class—and receive no employee benefits. Especially in the liberal arts, these adjunct positions have become dead-end jobs for huge numbers of PhD graduates who once hoped for tenure-track academic careers.

While instructional costs have been largely controlled, the amount spent on administration and facilities has soared. At many large campuses, the number of administrators now exceeds the number of instructors. During the same two-year period in which faculty employment fell by over 2 percent at the University of California, jobs for managers increased by 4.2 percent. Spending on professionals offering personalized counseling and advice to students likewise soared, and positions of this type now constitute almost a third of professional jobs at major American universities.[20] The higher-education industry has seemingly become a self-perpetuating jobs machine for the highly credentialed—unless, that is, you actually want a job teaching. The other major money pit has been extraordinarily zealous investment in luxurious student housing, recreation, and sports facilities. Selingo notes that "the most absurd frill is the Lazy River, essentially a theme park water ride where students float on rafts."[21] Administrators at Boston University, the University of Akron, the University of Alabama, and the University of Missouri all deem this an indispensable part of the college experience.

The most important factor, of course, has simply been the willingness of students and their families to pay ever-higher prices for an essential—if not sufficient—ticket into the middle class. Small wonder, then, that many observers have expressed the view that higher education has become a "bubble," or at least a bloated house of cards that is ripe for the same kind of digital decimation that has already transformed the newspaper and magazine industries. MOOCs offered by elite institutions are viewed as the mechanism mostly likely to impose the winner-take-all scenario that invariably takes hold once an industry goes digital.

The United States has over 2,000 four-year colleges and universities. If you include institutions that grant two-year degrees, the number grows to over 4,000. Of these, perhaps 200–300 might be characterized as selective. The number of schools with national reputations, or that might be considered truly elite, is, of course, far smaller. Imagine a future where college students can attend free online courses taught by Harvard or Stanford professors and subsequently receive a credential that would be acceptable to employers or graduate schools. Who, then, would be willing to go into debt in order to pay the tuition at a third- or fourth-tier institution?

Clayton Christensen, a professor at Harvard Business School and an expert in disruptive innovation within industries, has predicted that the answer to that question will result in a grim future for thousands of institutions. In a 2013 interview, Christensen said that "15 years from now, half of US universities may be in bankruptcy."[22] Even if most institutions remain solvent, it is easy to imagine dramatically declining enrollments and revenues coupled with massive layoffs of both administrators and faculty.

Many people assume that the disruption will come from the very top, as students flock to courses offered by Ivy League institutions. Yet, this assumes that "education" is the primary product that will be digitized. The very fact that schools like Harvard and Stanford are willing to give that education away for free is evidence that these institutions are primarily in the business of conveying credentials rather than knowledge. Elite credentials do not scale in the same way as, say, a digital music file; they are more like limited-edition art prints or paper money created by a central bank. Give away too many and their value falls. For this reason, I suspect that truly top-tier colleges will remain quite wary of providing meaningful credentials.

The disruption may be more likely to come from the next tier, especially major public universities that have strong academic reputations and huge numbers of alumni—as well as brands anchored by high-profile football and basketball programs—and are increasingly desperate for revenue in the wake of state funding cuts. Georgia Tech's partnership with Udacity to offer a MOOC-based computer science degree and the University of Wisconsin's experiment with competency-based credentials may offer previews of what is soon to arrive on a far more massive scale. As I suggested previously, there might also be opportunities for one or more private firms to claim a large slice of the market by offering vocationally oriented credentials based purely on assessment tests.

Even if MOOCs don't soon evolve into a direct path to a degree or other marketable credential, they could still undermine the business models of many colleges on a class-by-class basis. Large introductory lectures in courses like economics and psychology are vital cash cows for colleges because they require relatively few resources to teach hundreds of students—most of whom are paying full freight. If students at some point have the option of substituting a free or low-cost MOOC taught by a celebrity professor at an elite institution, that alone could be a major blow to the financial stability of many lower-ranked schools.

As MOOCs continue to evolve, their huge enrollments will themselves be an important driver of innovation. A vast volume of data is being collected about the students who participate and the ways in which they succeed or fail as they proceed through the courses. As we've seen, big data techniques are sure to result in important insights that will lead to improved outcomes over time. New educational technologies are also emerging and will increasingly be incorporated into MOOCs. Adaptive learning systems, for example, provide what amounts to a robotic tutor. These systems closely follow the progress of individual students and offer personalized instruction and assistance. They can also adjust the pace of learning to match student capabilities. Such systems are already proving to be successful. One randomized study looked at introductory statistics courses at six public universities. The students in one group took the course in a traditional format, while those in the other received primarily robotic instruction combined with limited classroom time. The study found that both groups of students performed at the same levels "in terms of pass rates, final exam scores, and performance on a standardized assessment of statistical literacy."[23]

If the higher-education industry ultimately succumbs to the digital onslaught, the transformation will very likely be a dual-edged sword. A college credential may well become less expensive and more accessible to many students, but at the same time, technology could devastate an industry that is itself a major nexus of employment for highly educated workers. And as we've already seen, in an entire range of other industries, advancing automation software will continue to impact many of the higher-skill jobs these new graduates are likely to seek. Even as essay-grading algorithms and robotic tutors help teach students to write, algorithms like those developed by Narrative Science might have already automated much of the routine, entry-level writing in many areas.

There may also prove to be a natural synergy between the rise of MOOCs and the practice of offshoring knowledge-based jobs. If massive online courses eventually lead to college degrees, it seems inevitable that a great many of the people—and a high percentage of the top-performing candidates—awarded these new credentials will be located in the developing world. As employers become accustomed to hiring workers educated via this new paradigm, they may also be inclined to take an increasingly global approach to recruiting.

HIGHER EDUCATION is one of two major US industries that has, so far, been relatively immune to the impact of accelerating digital technology. Nonetheless, innovations like MOOCs, automated grading algorithms, and adaptive learning systems offer a relatively promising path toward eventual disruption. As we'll see next, the other major holdout—health care—represents an even greater challenge for the robots.

Chapter 6

THE HEALTH CARE CHALLENGE

In May 2012, a fifty-five-year-old man checked into a clinic at the University of Marburg in Germany. The patient suffered from fever, an inflamed esophagus, low thyroid hormone levels, and failing vision. He had visited a series of doctors, all of whom were baffled by his condition. By the time he arrived at the Marburg clinic, he was nearly blind and was on the verge of heart failure. Months earlier, and a continent away, a very similar medical mystery had culminated with a fifty-nine-year-old woman receiving a heart transplant at the University of Colorado Medical Center in Denver.

The answer to both mysteries turned out to be the same: cobalt poisoning.[1] Both patients had previously received artificial hips made from metal. The metal implants had abraded over time, releasing cobalt particles and exposing the patients to chronic toxicity. In a remarkable coincidence, papers describing the two cases were published independently in two leading medical journals on nearly the same day in February 2014. The report published by the German doctors came with a fascinating twist: whereas the American team had resorted to surgery, the German team had managed to solve the mystery not because of their training but because one of the doctors had seen a February 2011 episode of the television show *House*. In the episode, the show's protagonist, Dr. Gregory House, is faced with the same problem and makes an ingenious diagnosis: cobalt poisoning resulting from a metal prosthetic hip replacement.

The fact that two teams of doctors can struggle to make the same diagnosis—and that they can do so even when the answer to the mystery has been broadcast to millions of prime-time television viewers—is a testament to the extent to which medical knowledge and diagnostic skill are compartmentalized in the brains of individual physicians, even in an age when the Internet has enabled an unprecedented degree of collaboration and access to information. As a result, the fundamental process that doctors use to diagnose and treat illnesses has remained, in important ways, relatively unchanged. Upending that traditional approach to problem solving, and unleashing all the information trapped in individual minds or published in obscure medical journals, likely represents one of the most important potential benefits of artificial intelligence and big data as applied to medicine.

In general, the advances in information technology that are disrupting other areas of the economy have so far made relatively few inroads into the health care sector. Especially hard to find is any evidence that technology is resulting in meaningful improvements in overall efficiency. In 1960, health care represented less than 6 percent of the US economy.[2] By 2013 it had nearly tripled, having grown to nearly 18 percent, and per capita health care spending in the United States had soared to a level roughly double that of most other industrialized countries. One of the greatest risks going forward is that technology will continue to impact asymmetrically, driving down wages or creating unemployment across most of the economy, even as the cost of health care continues to climb. The danger, in a sense, is not too many health care robots but too few. If technology fails to rise to the health care challenge, the result is likely to be a soaring, and ultimately unsustainable, burden on both individual households and the economy as a whole.

Artificial Intelligence in Medicine

The total amount of information that could potentially be useful to a physician attempting to diagnose a particular patient's condition or design an optimal treatment strategy is staggering. Physicians are faced with a continuous torrent of new discoveries, innovative treatments, and clinical study evaluations published in medical and scientific journals throughout the world. For example, MEDLINE, an online database maintained by the US National Library of Medicine, indexes over 5,600 separate journals—each of which might publish anywhere from dozens to hundreds of distinct research papers every year. In addition, there are millions of medical records, patient histories, and case studies that might offer important insights. According to one estimate, the total volume of all this data doubles roughly every five years.[3] It would be impossible for any human being to assimilate more than a tiny fraction of the relevant information even within highly specific areas of medical practice.

As we saw in Chapter 4, medicine is one of the primary areas where IBM foresees its Watson technology having a transformative impact. IBM's system is capable of churning through vast troves of information in disparate formats and then almost instantly constructing inferences that might elude even the most attentive human researcher. It's easy to imagine a near-term future where such a diagnostic tool is considered indispensable, at least for physicians confronting especially challenging cases.

The MD Anderson Cancer Center at the University of Texas handles over 100,000 patients at its Houston hospital each year and is generally regarded as the best cancer treatment facility in the United States. In 2011, IBM's Watson team began working with MD Anderson's doctors to build a customized version of the system geared toward assisting oncologists working with leukemia cases. The goal is to create an interactive adviser capable of recommending the best evidence-based treatment options, matching patients with clinical drug trials, and highlighting possible dangers or side effects that might threaten specific patients. Initial progress on the project proved to be somewhat slower than the team expected, largely because of the challenges associated with designing algorithms capable of taking on the complexities of cancer diagnosis and treatment. Cancer, it turns out, is tougher than *Jeopardy!* Nonetheless, by January 2014, the *Wall Street Journal* reported that the Watson-based leukemia system at MD Anderson was "back on track" toward becoming operational.[4] Researchers hope to expand the system to handle other kinds of cancer within roughly two years. It's very likely that the lessons IBM takes away from this pilot program will enable the company to streamline future implementations of the Watson technology.

Once the system is operating smoothly, the MD Anderson staff plans to make it available via the Internet so that it can become a powerful resource for doctors everywhere. According to Dr. Courtney DiNardo, a leukemia expert, the Watson technology has the "potential to democratize cancer care" by allowing any physician to "access the latest scientific knowledge and MD Anderson's expertise." "For physicians who aren't leukemia experts," she added, the system "can function as an expert second opinion, allowing them to access the same knowledge and information" relied on by the nation's top cancer treatment center. DiNardo also believes that, beyond offering advice for specific patients, the system "will provide an unparalleled research platform that can be used to generate questions, explore hypotheses and provide answers to critical research questions."[5]

Watson is currently the most ambitious and prominent application of artificial intelligence to medicine, but there are other important success stories as well. In 2009, researchers at the Mayo Clinic in Rochester, Minnesota, built an artificial neural network designed to diagnose cases of endocarditis—an inflammation of the inner layer of the heart. Endocarditis normally requires that a probe be inserted into the patient's esophagus in order to determine whether or not the inflammation is caused by a potentially deadly infection—a procedure that is uncomfortable, expensive, and itself carries risks for the patient. The Mayo doctors instead trained a neural network to make the diagnosis based on routine tests and observable symptoms alone, without the need for the invasive technique. A study involving 189 patients found that the system was accurate more than 99 percent of the time and successfully saved over half of the patients from having to needlessly undergo the invasive diagnostic procedure.[6]

One of the most important benefits of artificial intelligence in medicine is likely to be the avoidance of potentially fatal errors in both diagnosis and treatment. In November 1994, Betsy Lehman, a thirty-nine-year-old mother of two and a widely read columnist who wrote about health-related issues for the *Boston Globe,* was scheduled to begin her third round of chemotherapy as she continued her battle against breast cancer. Lehman was admitted to the Dana-Farber Cancer Institute in Boston, which, like MD Anderson, is regarded as one of the country's preeminent cancer centers. The treatment plan called for Lehman to be given a powerful dose of cyclophosphamide—a highly toxic drug intended to wipe out her cancer cells. The research fellow who wrote the medication order made a simple numerical error, which meant that the total dosage Lehman received was about four times what the treatment plan actually called for. Lehman died from the overdose on December 3, 1994.[7]

Lehman was just one of as many as 98,000 patients who die in the United States each year as a direct result of preventable medical errors.[8] A 2006 report by the US Institute of Medicine estimated that at least 1.5 million Americans are harmed by medication errors alone, and that such mistakes result in more than $3.5 billion in additional annual treatment costs.[9] An AI system with access to detailed patient histories, as well as information about medications, including their associated toxicity and side effects, would potentially be able to prevent errors even in very complex situations involving the interaction of multiple drugs. Such a system could act as an interactive adviser to doctors and nurses, offering instantaneous verification of both safety and effectiveness before medication is administered, and—especially in situations where hospital staff are tired or distracted—it would be very likely to save both lives and needless discomfort and expense.

Once medical applications of artificial intelligence evolve to the point where the systems can act as true advisers capable of providing consistently high-quality second opinions, the technology could also help rein in the high costs associated with malpractice liability. Many physicians feel the need to practice "defensive medicine" and order every conceivable test in an attempt to protect themselves against potential lawsuits. A documented second opinion from an AI system versed in best practice standards could offer doctors a "safe harbor" defense against such claims. The result might be less spending on needless medical tests and scans as well as lower malpractice insurance premiums.*

Looking even further ahead, we can easily imagine artificial intelligence having a genuinely transformative impact on the way medical services are delivered. Once machines demonstrate that they can offer accurate diagnosis and effective treatment, perhaps it will not be necessary for a physician to directly oversee every encounter with every patient.

---

* This raises the question of whether the liability would simply migrate to the manufacturer of the AI system. Since such systems might be used to diagnose tens or even hundreds of thousands of patients, the potential liability for errors could be daunting. However, the US Supreme Court ruled in the 2008 case *Riegel v. Medtronic, Inc.*, that medical device manufacturers are protected from some lawsuits if their products have been approved by the FDA. Perhaps similar reasoning would be extended to diagnostic systems. Another issue is that previous attempts to create "safe harbor" laws for doctors have been vigorously opposed by the trial lawyers, who have a great deal of political influence.

In an op-ed I wrote for the *Washington Post,* shortly after Watson's 2011 triumph at playing *Jeopardy!*, I suggested that there may eventually be an opportunity to create a new class of medical professionals: persons educated with perhaps a four-year college or master's degree, and who are trained primarily to interact with and examine patients—and then to convey that information into a standardized diagnostic and treatment system.[10] These new, lower-cost practitioners would be able to take on many routine cases, and could be deployed to help manage the dramatically growing number of patients with chronic conditions such as obesity and diabetes.

Physicians groups would, of course, be likely to oppose the influx of these less-educated competitors.* However, the reality is that the vast majority of medical school graduates are not especially interested in entering family practice, and they are even less excited about serving rural areas of the country. Various studies predict a shortage of up to 200,000 doctors within the next fifteen years as older doctors retire, the Affordable Care Act plan brings as many as 32 million new patients into the health insurance system, and an aging population requires more care.[11] The shortage will be most acute among primary-care physicians as medical school graduates, typically burdened by onerous levels of student debt, choose overwhelmingly to enter more lucrative specialties.

These new practitioners, trained to utilize a standardized AI system that encapsulates much of the knowledge that doctors acquire during the course of nearly a decade of intensive training, could handle routine cases, while referring patients who require more specialized care to physicians. College graduates would benefit significantly from the availability of a compelling new career path, especially as intelligent software increasingly erodes opportunities in other sectors of the job market.

---

* Nurse practitioners with advanced degrees have been able to overcome such political opposition in seventeen US states and are likely to be an important component of primary care in the future.

In some areas of medicine, particularly those that don't require direct interaction with patients, advances in AI are poised to drive dramatic productivity increases and perhaps eventually full automation. Radiologists, for example, are trained to interpret the images that result from various medical scans. Image processing and recognition technology is advancing rapidly and may soon be able to usurp the radiologist's traditional role. Software can already recognize people in photos posted on Facebook and even help identify potential terrorists in airports. In September 2012, the FDA approved an automated ultrasound system for screening women for breast cancer. The device, designed by U-Systems, Inc., is designed to help identify cancer in the roughly 40 percent of women whose dense breast tissue can render standard mammogram technology ineffective. Radiologists still need to interpret the images, but doing so now takes only about three minutes. That compares with twenty to thirty minutes for images produced using standard handheld ultrasound technology.[12]

Automated systems can also provide a viable second opinion. A very effective—but expensive—way to increase cancer detection rates is to have two radiologists read every mammogram image separately and then reach a consensus on any potential anomalies identified by either doctor. This "double reading" strategy results in significantly improved cancer detection and also dramatically reduces the number of patients who have to be recalled for further testing. A 2008 study published in the *New England Journal of Medicine* found that a machine can step into the role of the second doctor. When a radiologist is paired with a computer-aided detection system, the results are just as good as having two doctors separately interpret the images.[13]

Pathology is another area where artificial intelligence is already encroaching. Each year, over a hundred million women throughout the world receive a Pap test to screen for cervical cancer. The test requires that cervical cells be deposited on a glass microscope slide and then be examined by a technician or doctor for signs of malignancy. It's a labor-intensive process that can cost up to $100 per test. Many diagnostic labs, however, are now turning to a powerful automated imaging system manufactured by BD, a New Jersey–based medical device company. In a 2011 series of articles about job automation for *Slate*, technology columnist Farhad Manjoo called the BD FocalPoint GS Imaging System "a marvel of medical engineering" whose "image-searching software rapidly scans slides in search of more than 100 visual signs of abnormal cells." The system then "ranks the slides according to the likelihood they contain disease" and finally "identifies 10 areas on each slide for a human to scrutinize."[14] The machine does a significantly better job of finding instances of cancer than human analysts alone, even as it roughly doubles the speed at which the tests can be processed.

Hospital and Pharmacy Robotics

The pharmacy at the University of California Medical Center in San Francisco prepares about 10,000 individual doses of medication every day, and yet a pharmacist never touches a pill or a medicine bottle. A massive automated system manages thousands of different drugs and handles everything from storing and retrieving bulk pharmaceutical supplies to dispensing and packaging individual tablets. A robotic arm continuously picks pills from an array of bins and places them in small plastic bags. Every dose goes into a separate bag and is labeled with a barcode that identifies both the medication and the patient who should receive it. The machine then arranges each patient's daily meds in the order that they need to be taken and binds them together. Later, the nurse who administers the medication will scan the barcodes on both the dosage bag and the patient's wrist band. If they don't match, or if the medication is being given at the wrong time, an alarm sounds. Three other specialized robots automate the preparation of injectable medicines; one of these robots deals exclusively with highly toxic chemotherapy drugs. The system virtually eliminates the possibility of human error by cutting humans almost entirely out of the loop.

UCSF's $7 million automated system is just one of the more spectacular examples of the robotic transformation that's unfolding in the pharmacy industry. Far less expensive robots, not much larger than a vending machine, are invading retail pharmacies located in drug and grocery stores. Pharmacists in the United States require extensive training (a four-year doctoral degree) and have to pass a challenging licensing exam. They are also well paid, earning about $117,000 on average in 2012. Yet, especially in retail settings, much of the work is fundamentally routine and repetitive, and the overriding concern is to avoid a potentially deadly mistake. In other words, much of what pharmacists do is almost ideally suited to automation.

Once a patient's medication is ready to leave a hospital pharmacy, it's increasingly likely that it will do so in the care of a delivery robot. Such machines already cruise the hallways in huge medical complexes delivering drugs, lab samples, patient meals, or fresh linens. The robots can navigate around obstacles and use elevators. In 2010, El Camino Hospital in Mountain View, California, leased nineteen delivery robots from Aethon, Inc., at an annual cost of about $350,000. According to one hospital administrator, paying people to do the same work would have cost over a million dollars per year.[15] In early 2013, General Electric announced plans to develop a mobile robot capable of locating, cleaning, sterilizing, and delivering the thousands of surgical tools used in operating rooms. The tools would be tagged with radio-frequency identification (RFID) locator chips, making it easy for the machine to find them.[16]

Beyond the specific areas of pharmacy and hospital logistics and delivery, autonomous robots have so far made relatively few inroads. Surgical robots are in widespread use, but they are designed to extend the capabilities of surgeons, and robotic surgery actually costs more than traditional methods. There is some preliminary work being done on building more ambitious surgical robots; for example, the I-Sur project is an EU-backed consortium of European researchers who are attempting to automate basic procedures like puncturing, cutting, and suturing.[17] Still, for the foreseeable future, it seems inconceivable that any patient would be allowed to undergo an invasive procedure without a doctor being present and ready to intervene, so even if such technology materializes, any cost savings would likely be marginal at best.

Elder-Care Robots

The populations of all advanced countries, as well as many developing nations, are aging rapidly. The United States is projected to have over 70 million senior citizens, making up about 19 percent of the population, by 2030. That's up from just 12.4 percent in 2000.[18] In Japan, longevity combined with a low birth rate make the problem even more extreme; by 2025 fully a third of the population will be over sixty-five. The Japanese also have a nearly xenophobic aversion to the increased immigration that might help mitigate the problem. As a result, Japan already has at least 700,000 fewer elder-care workers than it needs—and the shortage is expected to become far more severe in the coming decades.[19]

This surging global demographic imbalance is creating one of the greatest opportunities in the field of robotics: the development of affordable machines that can assist in caring for the elderly. The 2012 movie *Robot & Frank*, a comedy that tells the story of an elderly man and his robotic caretaker, offers a very hopeful take on the kind of progress we're likely to see. The movie opens by announcing to the viewer that it is set in the "near future." The robot then proceeds to exhibit extraordinary dexterity, carry out intelligent conversations, and generally act just like a person. At one point, a glass is knocked off a table, and the robot snatches it out of midair. That, I'm afraid, is not a "near future" scenario.

Indeed, the main problem with elder-care robots as they exist today is that they really don't do a whole lot. Much of the initial progress has been with therapeutic pets like Paro, a robotic baby seal that provides companionship (at a cost of up to $5,000). Other robots are able to lift and move elderly people, saving a great deal of wear and tear on human caretakers. However, such machines are expensive and heavy—they may weigh ten times as much as the person they are lifting—and will, therefore, probably be deployed primarily in nursing homes or hospitals. Building a low-cost robot with sufficient dexterity to assist with personal hygiene or using the bathroom remains an extraordinary challenge. Experimental machines capable of specific tasks have appeared. For example, researchers at Georgia Tech have built a robot with a soft touch that can give patients a gentle bed bath, but the realization of an affordable, multitasking elder-care robot that can autonomously assist people who are almost completely dependent on others probably remains far in the future.

One of the ramifications of that daunting technical hurdle is that, despite the theoretically huge market opportunity, there are relatively few start-up companies focused on designing elder-care robots and little venture capital flowing into the field. The best hope almost certainly comes from Japan, which is on the brink of a national crisis and which, unlike the United States, has little aversion to direct collaboration between industry and government. In 2013, the Japanese government initiated a program in which it will pay two-thirds of the costs associated with developing inexpensive, single-task robotic devices that can assist the elderly or their caretakers.[20]

Perhaps the most remarkable elder-care innovation developed in Japan so far is the Hybrid Assistive Limb (HAL)—a powered exoskeleton suit straight out of science fiction. Developed by Professor Yoshiyuki Sankai of the University of Tsukuba, the HAL suit is the result of twenty years of research and development. Sensors in the suit are able to detect and interpret signals from the brain. When the person wearing the battery-powered suit thinks about standing up or walking, powerful motors instantly spring into action, providing mechanical assistance. A version is also available for the upper body and could assist caretakers in lifting the elderly.

Wheelchair-bound seniors have been able to stand up and walk with the help of HAL. Sankai's company, Cyberdyne, has also designed a more robust version of the exoskeleton for use by workers cleaning up the Fukushima Daiichi nuclear plant in the wake of the 2011 disaster. The company says the suit will almost completely offset the burden of over 130 pounds of tungsten radiation shielding worn by workers.* HAL is the first elder-care robotic device to be certified by Japan's Ministry of Economy, Trade, and Industry. The suits lease for just under $2,000 per year and are already in use at over three hundred Japanese hospitals and nursing homes.[21]

Other near-term developments will probably include robotic walkers to assist in mobility and inexpensive robots capable of bringing medicine, providing a glass of water, or retrieving commonly misplaced items like eyeglasses. (This would likely be done by attaching RFID tags to the items.) Robots that can help track and monitor people with dementia are also appearing. Telepresence robots that allow doctors or caretakers to interact with patients remotely are already in use in some hospitals and care facilities. Devices of this type are relatively easy to develop because they skirt around the challenge of dexterity. The near-term nursing-care robotics story is primarily going to be about machines that assist, monitor, or enable communication. Affordable robots that can independently perform genuinely useful tasks will be slower to arrive.

Given that truly capable and autonomous elder-care robots are unlikely to emerge in the near future, it might seem reasonable to expect that the looming shortage of nursing home workers and home health aids will, to a significant extent, offset any technology-driven

* The names selected by Sankai seem a bit odd for a company focused primarily on elder care. HAL, of course, was the unfriendly computer that wouldn't open the pod bay doors in *2001: A Space Odyssey.* Cyberdyne was the fictional corporation that built Skynet in the *Terminator* movies. Perhaps the company is eying other markets.

job losses that occur in other sectors of the economy. Maybe employment will simply migrate to the health and elder care sector. The US Bureau of Labor Statistics (BLS) projects that by 2022, there will be 580,000 new jobs for personal-care aids and 527,000 for registered nurses (those are the two fastest-growing occupations in the United States), as well as 424,000 home heath aids and 312,000 nursing aids.[22] That adds up to about 1.8 million jobs.

This sounds like a big number. But now consider that the Economic Policy Institute estimates that, as of January 2014, the United States was still short 7.9 million jobs as a result of the Great Recession. That includes 1.3 million jobs that were lost during the downturn and hadn't yet been recovered as well as another 6.6 million jobs that were never created.[23] In other words, if those 1.8 million jobs all appeared today, they would fill only about a quarter of the hole.

Another factor, of course, is that these jobs are low-paying and not particularly suitable for a large fraction of the population. According to the BLS, home health aids and personal aids both provided a medium 2012 income of under $21,000 and require an education level of "less than high school." Large numbers of workers are likely to lack the temperament necessary to thrive in these jobs. If a worker hates his job stamping out widgets, that's one thing. If he despises his job caring for a dependent older person, that's a major problem.

Assuming the BLS's projections are correct and these jobs do materialize in large numbers, there is also the question of who will actually pay for these workers. Decades of stagnant wages, together with the transition from defined benefit pensions to often under-funded 401k plans, will leave a large fraction of Americans in relatively insecure retirement situations. By the time the majority of older people reach the point where they need personal, daily assistance, relatively few are likely to have the private means to hire home health aids, even if the wages for these jobs continue to be very low. As a result, these will probably be quasi-government jobs funded by programs like Medicare or Medicaid and will therefore be viewed as more of a problem than a solution.

Unleashing the Power of Data

As we saw in Chapter 4, the big data revolution offers the promise of new management insights and significantly improved efficiency. In fact, the increasing importance of all this data may be a powerful argument for consolidation in the health insurance sector, or alternatively creating some mechanism for sharing data among insurance companies, hospitals, and other providers. Access to more data could well mean more innovation. Just as Target, Inc., was able to predict pregnancy based on customer purchasing patterns, hospitals or insurance companies with access to large datasets will potentially discover correlations between specific factors that can be controlled and the likelihood of a positive patient outcome. The original AT&T was famous for sponsoring Bell Labs, where many of the twentieth century's most important advances in information technology took place. Perhaps one or more health insurance companies with sufficient scale could play a somewhat similar role—except that the innovations would come not from tinkering in a lab but from continuously analyzing reams of detailed patient and hospital operational data.

Medical sensors either implanted or attached to patients will provide another important source of data. These devices will produce a continuous stream of biometric information that can be used in both diagnosis and in the management of chronic diseases. One of the most promising areas of research is the design of sensors capable of monitoring glucose in people with diabetes. The sensors could communicate with a smart phone or other external device, instantly alerting patients if their glucose level falls outside the safe range and avoiding the need for uncomfortable blood tests. A number of companies already manufacture glucose monitors that can be embedded under a patient's skin. In January 2014, Google announced that it is working on a contact lens that would contain a tiny glucose detector and wireless chip. The lenses would continuously monitor glucose levels by analyzing tears; if the wearer's blood sugar is too high or too low, a tiny LED light would illuminate, providing an instant alert.

Consumer devices like the Apple Watch, formally announced in September 2014, will likewise result in a torrent of health-related data.

Health Care Costs and a Dysfunctional Market

The March 4, 2013, issue of *Time* magazine featured a cover story by Steven Brill entitled "Bitter Pill." The article delved into the forces underlying ever-escalating health care costs in the United States and highlighted case after case of what can only be categorized as outright price gouging—including, for example, a 10,000 percent markup on the same over-the-counter acetaminophen tablets you could buy at your local drug store or Walmart. Routine blood tests for which Medicare would pay about $14 were marked up to $200 and beyond. CT scans that Medicare prices at about $800 were inflated to over $6,500. A feared heart attack that turned out to be a case of heartburn resulted in a $17,000 charge—not including fees for the doctor.[24]

A few months later, Elisabeth Rosenthal of the *New York Times* wrote a series of articles telling essentially the same story: a laceration requiring three simple stitches came in at well over $2,000. A dab of skin glue on a toddler's forehead cost over $1,600. One patient was charged nearly $80 for a small bottle of local anesthetic that can be purchased for $5 on the Internet. Rosenthal noted that the hospital, which buys such supplies in bulk, would likely pay far less.[25]

Both reporters found that these inflated charges generally originate with a massive, obscure—and often secretive—list of prices known as the "chargemaster." The prices listed in the chargemaster seemingly have no rhyme or reason and no meaningful relationship to actual costs. The only thing one can say with consistent certainty about the chargemaster is that its prices are very, very high. Both Brill and Rosenthal found that the most egregious cases of chargemaster abuse occurred with uninsured patients. Hospitals typically expected these people to pay full list price and often were quick to hire bill collectors or even file lawsuits if patients couldn't or wouldn't pay. Even major health insurance companies, however, are increasingly billed at rates based on a discount from chargemaster prices. In other words, the costs are first inflated—in many cases by a factor of ten or even a hundred—and then a discount of perhaps 30, or even 50, percent is applied, depending on how effectively the insurer negotiates. Imagine buying a gallon of milk for $20 after negotiating a 50 percent discount from the $40 list price. Given this, it should come as no surprise that hospital charges are the most important single driver of consistently soaring health care costs in the United States.

One of the most important lessons of history is that there is a powerful symbiosis between technological progress and a well-functioning market economy. Healthy markets create the incentives that lead to meaningful innovation and ever-increasing productivity, and this has been the driving force behind our prosperity.* Most intelligent people understand this (and are very likely to bring up Steve Jobs and the iPhone when discussing it). The problem is that health care is a broken market and no amount of technology is likely to bring down costs unless the structural problems in the industry are resolved.

There is also, I think, a great deal of confusion about the nature of the health care market and exactly where an effective market pricing mechanism should come into play. Many people would like to believe that health care is a normal consumer market: if only we could get insurance companies, and especially the government, out of the way and instead push decisions and costs onto the consumer (or patient), then we'd get innovations and outcomes similar to what we've seen in other industries (Steve Jobs might be mentioned again here).

---

* Consider, for example, the Soviet Union, which by all accounts had some of the best scientists and engineers in the world. The Soviets were able to achieve solid results in military and space technology, but they were never able to scale the benefits of innovation across the civilian economy. The reason certainly has a lot to do with the absence of working markets.

The reality, however, is that health care is simply not comparable to other markets for consumer products and services, and this has been well understood for over half a century. In 1963, the Nobel laureate economist Kenneth Arrow wrote a paper detailing the ways in which medical care stands apart from other goods and services. Among other things, Arrow's paper highlighted the fact that medical costs are extremely unpredictable and often very high, so that consumers can neither pay for them out of ongoing income nor effectively plan ahead as they might for other major purchases. Medical care can't be tested before you buy it; it's not like visiting the wireless store and trying out all the smart phones. In emergencies, of course, the patient may be unconscious or about to die. And, in any case, the whole business is so complex and requires so much specialized knowledge that a normal person can't reasonably be expected to make such decisions. Health care providers and patients simply don't come to the table as anything approaching equals, and as Arrow pointed out, "both parties are aware of this informational inequality, and their relation is colored by this knowledge."[26] The bottom line is that the high cost, unpredictability, and complexity of major medical and hospitalization services make some kind of insurance model essential for the health care industry.

It is also critical to understand that health care spending is highly concentrated among a tiny number of very sick people. A 2012 report by the National Institute for Health Care Management found that just 1 percent of the population—the very sickest people—accounted for over 20 percent of total national health care spending. Nearly *half* of all spending, about $623 billion in 2009, went to the sickest 5 percent of the population.[27] In fact, health care spending is subject to the same kind of inequality as income in the United States. If you draw a graph, it will look very much like the winner-take-all/long-tail distribution I described in Chapter 3.

The importance of this intense concentration of spending cannot be overemphasized. The small population of very ill people on whom we are spending all this money are obviously not in a position to negotiate prices with providers; nor would we want to place such a staggering fiscal responsibility in these people's hands. The "market" that we need to make work exists between the providers and the insurance companies—not between providers and patients. The essential lesson of the articles written by Brill and Rosenthal is that this market is dysfunctional because of a fundamental power imbalance between insurers and providers. While individual consumers may rightly perceive health insurance companies as powerful and domineering, the reality is that—relative to providers like hospitals, doctors, and the pharmaceutical industry—they are, in a great many cases, *too weak*. That imbalance is being steadily worsened by an ongoing wave of consolidations among providers. Brill's article notes that as hospitals increasingly snap up "doctor's practices and competing hospitals, their leverage over insurance companies is increasing."[28]

Imagine a near future where a physician wields a powerful tablet computer that allows her to order a range of medical tests and scans with just a few presses on her touch screen. Once a test is completed, the results are instantly routed to her device. If a patient needs a CT scan, or perhaps an MRI, the results are accompanied by a detailed analysis performed by an artificial intelligence application. The software points out any anomalies in the scan and makes recommendations for further care by accessing a massive database of patient records and identifying similar cases. The doctor can see exactly how comparable patients were treated, any issues that arose, and how things ultimately turned out. All this would, of course, be efficient and convenient and ought to lead to a better outcome for the patient. This is the kind of scenario that gets techno-optimists excited about the revolution soon to unfold in the health care arena.

Now assume that the doctor has a financial interest in the diagnostic company that performs the tests or scans. Or, then again, maybe the hospital has acquired the doctor's practice and also owns the testing facility. The prices for the tests and scans bear little relation to the actual costs of these services—after all, they're listed in the chargemaster—and they are highly profitable. Every time our doctor presses her touch screen, she essentially mints money.

While this example is, at the moment, imaginary, there is an abundance of evidence demonstrating that new health care technologies very often lead to more spending rather than improved productivity. The primary reason is that there is no effective market pricing mechanism to drive increased efficiency. In the absence of market pressure, providers often invest in technologies designed to increase revenue rather than efficiency, or where they do achieve increased productivity they simply retain the profits rather than lowering prices.

The poster child for technology investment as a driver of health care inflation may well be the "proton beam" facilities that are being built to treat prostate cancer. A May 2013 article by Jenny Gold of *Kaiser Health News* noted that "despite efforts to get health care spending under control, hospitals are still racing to build expensive new technology—even when the devices don't necessarily work better than the cheaper kind."[29] The article describes one proton beam facility as "a giant cement-encased building the size of a football field, with a price tag of more than $200 million." The idea behind this expensive new technology is that it delivers less radiation to patients, and yet, studies have found no evidence that protein beam technology results in better patient outcomes than far less expensive approaches.[30] Health care expert Dr. Ezekiel Emanuel says, "We don't have evidence that there's a need for them in terms of medical care. They're simply done to generate profits."[31]

To me, it seems evident that the American people could in principle be made much better off by a massive technological disruption of the health care sector than of, say, the fast food industry. After all, lower prices and improved productivity in health care will likely lead directly to better and longer lives. Cheaper fast food may well do the opposite. Yet, the fast food industry has well-functioning markets—and the health care sector does not. As long as that situation is allowed to persist, there are few reasons to be optimistic that accelerating technology alone will succeed in reining in soaring health care costs. Given this reality, I'd like to take a brief detour from our technology narrative in order to suggest two alternate strategies that might help to correct the power imbalance between insurers and providers, and hopefully enable the kind of synergy between markets and technology that might bring the transformation we hope for.

Consolidate the Industry and Treat Health Insurance as a Utility

One of the primary messages that leaps out from an analysis of the prices charged by providers is that Medicare—the government-run program for people aged sixty-five and over—is by far the most efficient portion of our health care system. As Brill writes, "Unless you are protected by Medicare, the health care market is not a market at all. It's a crapshoot." The implementation of the Affordable Care Act (Obamacare) will certainly improve the situation as far as individuals who previously lacked insurance are concerned, but it does relatively little to actively rein in hospital costs; instead, the inflated costs will be shifted to insurers and then ultimately to taxpayers in the form of the subsidies that were put in place to make health insurance affordable to people with moderate incomes.

The fact that Medicare is relatively effective at controlling most patient-related costs, while spending far less than private insurers on administration and overhead, underlies the argument for simply expanding the program to include everyone and, in effect, creating a single-payer system. This has been the path followed by a number of other advanced countries—all of which spend far less on health care than the United States and typically have better outcomes according to metrics like life expectancy and infant mortality. While a single-payer system, managed by the government, has both logic and evidence to support it, there is no escaping the reality that in the United States the whole idea is ideologically toxic to roughly half the population. Putting such a system in place would also presumably result in the demise of nearly the entire private health insurance sector; that does not seem likely given the enormous political influence wielded by the industry.

A single-payer system is, in practice, always assumed to be run by the government, but in theory this does not have to be the case. Another approach might be to merge all private insurance companies into a single national corporation, which would then be heavily regulated. The model would be the original AT&T before it was broken up in the 1980s. The central idea here is that health care is in many ways akin to the telecommunications system: it is, in essence, a utility. Like water and sanitation systems or the nation's electrical infrastructure, the health care system does not stand alone—it is a systemic industry whose efficient operation is critical to both the economy and society. In many cases, the provision of a utility service leads to natural monopoly scenarios. In other words, it is most efficient if only a single firm operates in the market.

An even more effective variation on this theme might be to allow a small number of large competing insurance companies—in effect, a sanctioned oligopoly. This would inject an element of competition into the system. The companies would still be large enough to have significant market power when negotiating with providers, and they would have little choice but to compete on the basis of enabling high-quality care since their reputations would determine their success. Tight regulation of the industry would limit price increases and prevent the companies from engaging in undesirable practices like, for example, designing insurance plans geared specifically toward "cherry-picking" younger, healthier patients or offering plans with substandard protection. Instead, they would have to focus on genuine innovation and efficiency.

Consolidating existing insurance companies into one or more regulated "health care utilities" might provide many of the advantages of a single-payer system while preserving the industry. Rather than being wiped out, the shareholders of private insurance companies might conceivably see gains as a result of an industry-wide merger. The mechanism by which such a consolidation might be brought about is, of course, far from obvious. Perhaps the government could issue a small number of operating licenses, and it might even hold an auction as it does for the electromagnetic communications spectrum.*

Set "All-Payer" Rates

An alternate, and perhaps more feasible, strategy is the implementation of an "all-payer" system. In this scenario, the government essentially sets the schedule of prices that can be charged by health care providers. Just as Medicare dictates the prices it will pay, an all-payer system would do the same for all patients receiving care from any given provider. An all-payer approach is used in the health care systems of a number of countries, including France, Germany, and Switzerland. In the United States, Maryland also has such a system for hospitals, and the state has seen relatively slow growth in hospitalization costs.[32] All-payer systems vary in the specifics of their implementation; the rates may be set through collective negotiation between providers and payers, or they might be established by a regulating commission after an analysis of actual costs at particular hospitals.

---

* In the United States, the constitutional authority to create a single-payer system—regardless of whether it is run by the government or by private corporations—probably derives from the government's ability to levy a tax on everyone to pay for the system. Therefore, all or a portion of the premiums would be paid by the government. This is already the case with the insurance subsidies associated with the Affordable Care Act. In other words, the federal government can force everyone to pay for a single-payer system through taxes, but it cannot prohibit a parallel private system. So there still would likely be additional services available to those willing and able to pay out of pocket, just as there are private schools. This is different from the system in Canada, where most private health care services are prohibited—leading some Canadians to seek health care services in the United States.

Since an all-payer system enforces the same prices for all patients, it has important implications for the cost shifting that goes on between private patients and those covered by the public systems in the United States (Medicaid for low-income people and Medicare for those over sixty-five). When a single rate is set, the public prices have to rise considerably, putting more of a burden on taxpayers. Privately insured patients, and especially those who are uninsured, will typically benefit from lower prices as they are no longer subsidizing the public programs. This has been the case with Maryland's program.*

It seems to me that a much simpler approach that might produce immediate savings would be to set an all-payer *ceiling* rather than a specific price. For instance, suppose the ceiling were set at the Medicare rate plus 50 percent. In one example from Brill's article, a blood test that Medicare says is worth $14 might then be priced at any amount up to $21—but it could never reach anything like $200. Insurance companies with sufficient market power would still be free to negotiate a price lower than the ceiling. This strategy would immediately eliminate the worst excesses, and as long as the ceiling was set high enough, it would still provide sufficient revenue to providers. A 2010 fact sheet published by the American Hospital Association claims that Medicare paid "90 cents for every dollar spent by hospitals caring for Medicare patients in 2009."[33] If the industry's own lobbying organization says Medicare is covering 90 percent of hospital costs, then a ceiling somewhat higher than the Medicare rate should be sufficient to allow enough cost shifting to make up for that missing 10 percent.** An all-payer ceiling would also be very

---

* Maryland has a special waiver that has been in place for over thirty years and allows it to pay higher Medicare rates. As of 2014, Maryland has moved to a new experimental system that is allowed under the Affordable Care Act. In addition to setting all-payer rates, the new program will enforce explicit caps on per capita hospital spending. The state expects to save $330 million in Medicare costs over a five-year period.

** The same fact sheet says that Medicaid (the program for the poor) paid 89 percent of actual hospital costs.

easy to implement since it is based directly on the already-published Medicare rates.

One of the most hopeful approaches to controlling health costs, which is gaining some traction in the current environment, is to transition away from a fee-for-service model and toward an "accountable care" system in which doctors and hospitals are paid a set fee to manage the overall health of patients. One of the primary advantages of this approach is that it would reorient the incentives regarding innovation. Rather than simply offering a new way to hoover up even higher fees according to a fixed schedule, emerging technologies would be viewed in terms of their potential to reduce costs and make care more efficient. The key to making that happen, however, is to push more of the financial risk associated with patient care away from insurers (or the government) and onto hospitals, doctors, and other providers. Needless to say, the latter are unlikely to accept that increased risk willingly. In other words, in order to drive a successful transition toward accountable care, we still need to address the market power imbalance that often exists between insurers and providers.

In order to bring relentlessly increasing health care costs in the United States under control, I think it will probably be necessary to pursue one of the two general strategies I've outlined. We will have to move toward a single-payer system where either the government or one or more large private firms exercise more bargaining power in the health insurance market, or alternatively we will need to have regulators exercise direct control over the rates paid to providers. In either scenario, moving aggressively toward an accountable care model might be a vital part of the solution. Both of these approaches, in various combinations, are used successfully by other advanced countries. The bottom line is that a pure "free market" approach in which we cut government out of the loop and expect patients to operate like consumers shopping for groceries or smart phones is never going to work. As Kenneth Arrow pointed out over fifty years ago, health care is simply different.

This is not to say that there are no significant dangers associated with either approach. Both strategies rely on regulators to either control premiums or set the prices paid to providers. There is an obvious risk of regulatory capture; powerful companies or industries may exert influence that bends government policy in their favor. Attempts at such influence have already been successfully directed at Medicare, which is specifically prohibited from using its market power to negotiate drug prices. The United States is virtually the only country in the world where this is the case; every other national government negotiates prices with the drug companies. The result is that Americans, in effect, subsidize lower drug prices in the rest of the world. The three years between 2006 and 2009 saw a 68 percent increase in the rate of "prescription abandonment" in the United States.[34] This happens when patients request that a prescription be filled, but then walk away when they find out the cost. It's something of a mystery to me why this is not more disturbing to Americans, and to grassroots conservatives in particular. The Tea Party, after all, got started after a famous rant by CNBC personality Rick Santelli, who decried the fact that people with mortgages they couldn't afford might be subsidized by taxpayers. Why aren't average Americans more upset about the fact that they are paying the pharmaceutical freight for the rest of the world—including a number of countries that have significantly higher per capita incomes than the United States?

In spite of this problem, Medicare consistently provides high-quality care at a cost significantly lower than in the highly fragmented private insurance sector. In other words, we should not make the perfect the enemy of the good. Nonetheless, Medicare's prohibition against negotiating with the pharmaceutical industry deserves to be subjected to a great deal more public scrutiny. The industry argues that inflated drug prices in the United States are necessary in order to fund further research. However, there are likely more efficient and certainly more equitable ways to ensure that drug research gets funded.[35*] The potential to reform or streamline the Federal Drug Administration's procedures for testing and approving new drugs also surely exists.

Another issue with Medicare, and one that touches directly on the subject of this book, is that waste can easily be driven by the direct advertisement of products to senior citizens who are told explicitly to pressure their physicians for a prescription and that Medicare will then pick up nearly the entire cost. One government audit found that up to 80 percent of the motorized scooters paid for by Medicare were not really needed by the elderly patients who received them and may actually be harmful to their health. The two largest scooter manufacturers spent over $180 million on advertisements directed at Medicare recipients in 2011.[36] This is another issue that deserves close scrutiny because, as we've seen, there is soon likely to be a profusion of robotic equipment geared toward providing home-based assistance to senior citizens. Such advances have great potential to improve quality of life for the elderly while reducing the cost of their care—but not if we pay for technology in cases where it is unneeded or perhaps even detrimental. The specter of millions of comfortably seated senior citizens watching advertisements telling them that Medicare will happily pay for a robot capable of retrieving their television remote should give us pause.**

---

* A related issue has to do with the patents granted to drug manufacturers. These prevent the introduction of cheaper, generic drugs for long periods. Many economists believe that the pharmaceutical patent system is very inefficient. Other countries can also potentially threaten to void drug patents as a price negotiating mechanism—putting a still higher burden on Americans. The Center for Economic and Policy Research published a briefing in 2004 that outlines these issues and presents some more efficient alternatives for funding drug research. Please see the corresponding endnote for details.

** The whole idea behind requiring prescriptions is that patients are not able (or cannot be trusted) to make these decisions for themselves. Why, then, do we allow drug companies or medical equipment manufacturers to advertise directly to patients?

WHILE RECENT APPLICATIONS OF AI and robotics to the health care field are impressive and advancing rapidly, they are, for the most part, just beginning to nibble at the edges of the hospital cost problem. With the exception of pharmacists, and possibly doctors or technicians who specialize in analyzing images or lab specimens, automating even a significant portion of the jobs done by most skilled health care workers remains a daunting challenge. For those seeking a career that is likely to be relatively safe from automation, a skilled health care profession that requires direct interaction with patients remains an excellent bet. That calculus could, of course, change in the more distant future. Twenty or thirty years from now, I think, it's impossible to say with any real confidence what might be technologically possible.

Technology is not the only consideration, of course. Health care, more than any other sector of the economy, is subject to a complex web of rules and regulations imposed by governments, agencies like the FDA, and licensing authorities. Every action and every decision are also colored by the looming threat of litigation if an error—or perhaps just an unlucky outcome—should occur. Even among retail pharmacists, the specific impact of automation on employment isn't easily discernible. The reason is likely regulation. Farhad Manjoo interviewed one pharmacist who said, "Most pharmacists are employed only because the law says that there has to be a pharmacist present to dispense drugs."[37] That, at least for the moment, is probably something of an exaggeration. Job prospects for newly minted pharmacists have worsened significantly over the past decade, and things may well get worse. A 2012 analysis identifies a "looming joblessness crisis for new pharmacy graduates" and suggests that the unemployment rate could reach 20 percent.[38] However, this is likely due largely to an explosion in the number of new graduates entering the job market as pharmacy schools have dramatically increased enrollments.* Relative

* One could also speculate that technology is *indirectly* contributing to diminished prospects for pharmacy graduates by driving more people into the profession. In the first decade of the new millennium, nearly fifty new pharmacy graduate schools opened their doors (a 60 percent increase), and existing programs also dramatically increased enrollments. The number of newly

to most other occupations, there's little doubt that health care professionals enjoy an extraordinary degree of employment security as a result of factors completely unrelated to the technical challenges associated with automating their jobs.

This may be good news for health care workers, but if technology has only a muted impact on health care costs even as it disrupts other employment sectors, the economic risks we face will be amplified. In that scenario, the burden of soaring health care costs will become even more unsustainable as advancing technology continues to produce unemployment and ever-increasing inequality, as well as stagnant, or even falling, incomes for most workers in other industries. This prospect makes it even more critical to introduce meaningful reforms that will correct the market power imbalance between insurers and providers so that advancing technology can be fully leveraged as a mechanism for increased efficiency across the health care sector. Without that, we run the risk that our market economy will eventually come to be dominated by a sector that is inefficient and, indeed, not an especially well-functioning market at all.

Controlling the health care cost burden is especially critical because, as we'll see in Chapter 8, the last thing American households need is an ever-increasing drain on their discretionary income. Indeed, stagnant incomes and growing inequality are already undermining the broad-based consumer demand that is vital to continued economic growth.

graduated pharmacists could hit 15,000 per year by 2016; that's over twice the number of degrees granted in 2000. Something very similar (and perhaps even more extreme) happened with law schools, and the law school enrollment bubble is now famously bursting. Law school has always been a well-traveled path toward monetizing a liberal arts degree. Pharmacy offers similar potential for an undergraduate biology degree. It may be that soaring demand for these professional degrees results, at least in part, from the evaporation of other good opportunities for college graduates. With relatively few other attractive alternatives, college graduates have clamored to get into law or pharmacy school, and the industry has responded by expanding enrollment and ultimately producing far more graduates than the market could absorb. The fact that both pharmacy and law are also impacted by direct automation makes things even more unsustainable. My prediction for the next professional school bubble: MBA degrees.

So far, we have focused primarily on the ways in which technology is likely to transform existing employment sectors. In the next chapter, we'll leap a decade or more ahead in time and imagine how things might look in a future economy populated with entirely new technologies and industries.

Chapter 7

TECHNOLOGIES AND INDUSTRIES OF THE FUTURE

YouTube was founded in 2005 by three people. Less than two years later, the company was purchased by Google for about $1.65 billion. At the time of its acquisition, YouTube employed a mere sixty-five people, the majority of them highly skilled engineers. That works out to a valuation of over $25 million per employee. In April 2012, Facebook acquired photo-sharing start-up Instagram for $1 billion. The company employed thirteen people. That's roughly $77 million per worker. Fast-forward another two years to February 2014 and Facebook once again stepped up to the plate, this time purchasing mobile messaging company WhatsApp for $19 billion. WhatsApp had a workforce of fifty-five—giving it a valuation of a staggering $345 million per employee.

Soaring per-employee valuations are a vivid demonstration of the way accelerating information and communications technology can leverage the efforts of a tiny workforce into enormous investment value and revenue. What's more, they offer compelling evidence for how the relationship between technology and employment has changed. There is a widely held belief—based on historical evidence stretching back at least as far as the industrial revolution—that while technology may certainly destroy jobs, businesses, and even entire industries, it will also create entirely new occupations, and the ongoing process of "creative destruction" will result in the emergence of new industries and employment sectors—often in areas that we can't yet imagine. A classic example is the rise of the automotive industry in the early twentieth century, and the corresponding demise of businesses engaged in manufacturing horse-drawn carriages.

As we saw in Chapter 3, however, information technology has now reached the point where it can be considered a true utility, much like electricity. It seems nearly inconceivable that successful new industries will emerge that do not take full advantage of that powerful new utility, as well as the distributed machine intelligence that accompanies it. As a result, emerging industries will rarely, if ever, be highly labor-intensive. The threat to overall employment is that as creative destruction unfolds, the "destruction" will fall primarily on labor-intensive businesses in traditional areas like retail and food preparation, while the "creation" will generate new businesses and industries that simply don't hire many people. In other words, the economy is likely on a path toward a tipping point where job creation will begin to fall consistently short of what is required to fully employ the workforce.

YouTube, Instagram, and WhatsApp are, of course, all examples drawn directly from the information technology sector, where we've come to expect tiny workforces and huge valuations and revenues. To illustrate how a similar phenomenon is likely to unfold on a much broader front, let's look in a bit more depth at two specific technologies that have the potential to loom large in the future: 3D printing and autonomous cars. Both are poised to have a significant impact within the next decade or so, and could eventually unleash a dramatic transformation in both the job market and the overall economy.

3D Printing

Three-dimensional printing, also known as additive manufacturing, employs a computer-controlled print head that fabricates solid objects by repeatedly depositing thin layers of material. This layer-by-layer construction method enables 3D printers to easily create objects with curves and hollows that might be difficult, or even impossible, to produce using traditional manufacturing techniques. Plastic is the most common construction material, but some machines can also print metal, as well as hundreds of other materials, including high-strength composites, flexible rubber-like substances, and even wood. The most sophisticated printers are able to build products containing as many as a dozen different materials. Perhaps most remarkably, the machines can print complex designs containing interlocking or moving parts as a single unit—eliminating any need for assembly.

A 3D printer lays down layers of material either by design or simply by copying an existing object using a 3D laser scanner or with sophisticated tools like computed tomography (CT scans). Late-night comedian Jay Leno, a classic-car enthusiast, uses this technique to produce replacement auto parts.

Three-dimensional printing is ideal for producing highly customized "one-off" products. The technology is already being used to build dental crowns, bone implants, and even prosthetic limbs. Design prototypes and architectural models are other popular applications.

An enormous amount of hype surrounds 3D printing and, in particular, its potential to upend the traditional factory-based manufacturing model. Much of this speculation is focused on the emergence of inexpensive desktop machines. Some enthusiasts foresee a future era of distributed fabrication, where virtually everyone owns a 3D printer and uses it to produce whatever he or she needs. Others project the rise of a new craft-based (or "maker") economy where small companies displace high-volume factory production with more personalized, locally produced products.

I think there are good reasons to be skeptical of such predictions. The most important reason is that the ease of customization offered by 3D printing comes at the cost of economies of scale. If you need to print a few copies of a document, you might do it on your home laser printer. If, however, you need 100,000 copies, it would be much more cost-effective to use a commercial printer. 3D printing versus traditional manufacturing involves essentially the same trade-off. While the printers themselves are rapidly falling in price, the same cannot be said of the material used in the process, especially if something other than plastic is required. The machines are also slow; building a substantial solid object in a consumer 3D printer can take several hours. Most of the products we use do not necessarily benefit from whole-scale customization; indeed, standardization often has important advantages. Three-dimensional printing might be a great way to create a custom case for your iPhone, but it seems very unlikely that you'll ever print the phone itself.*

If cheap desktop printers do become ubiquitous, that would likely destroy the market for finished products created with such machines. Instead, any value would reside entirely in the product's digital design file. Some entrepreneurs would be successful selling such designs, but the market would almost certainly evolve into the same winner-take-all scenario that characterizes other digital products and services. There would also be a multitude of free or open source

---

* Three-dimensional printers can already print basic electronic circuits, but it seems highly unlikely that they would ever be able to print the state-of-the-art processor and memory chips used in smart phones. Fabrication of these chips happens at industrial scale and requires precision vastly beyond the capability of any printer. One obvious future trend is that more and more of the everyday objects we use are likely to incorporate advanced processors and smart software. To me, this suggests that personal 3D printing is unlikely to keep pace with the products consumers really want to buy. A hobbyist, of course, might print most of a product and then assemble the necessary components, but I doubt that would appeal to most people.

designs—probably for nearly any conceivable product—available for download. The bottom line is that personal 3D printing would come to look much like the Internet: lots of free or inexpensive stuff for consumers, but far fewer opportunities for the vast majority of people to generate a significant income.

This is not to say that 3D printing won't be a transformative technology. The real action is likely to happen at industrial scale. Rather than displacing traditional manufacturing, 3D printing will be integrated with it. In fact, that's already happening. The technology has made significant inroads in the aerospace industry, where it is often used to create lighter-weight components. General Electric's aviation division plans to use 3D printing to produce at least 100,000 parts by 2020, resulting in a potential weight reduction of 1,000 pounds for a single aircraft engine.[1] To get a sense of how much fuel lopping half a ton off every engine could save, consider that in 2013, American Airlines replaced the paper flight manuals carried in its cockpits with digital versions loaded onto Apple iPads. That saved about 35 pounds per plane—and $12 million in annual fuel costs.[2] Cutting each plane's weight by an average of 3,000 pounds could save a billion dollars or more per year. One of the components that GE plans to print, a fuel nozzle, normally requires the assembly of twenty separate parts. A 3D printer will allow the entire component to be printed in one unit, fully assembled.[3]

As we saw in Chapter 1, manufacturing is likely to become more flexible, and in many cases, factories will be located closer to consumer markets. Three-dimensional printing will have a role to play in this transition. The technology will be used where it is most cost-effective: for example, in creating those parts that need to be customized, or perhaps in printing complex components that would otherwise require extensive assembly. Where 3D printing can't be used to directly fabricate high-volume parts, it will often find a role in rapidly creating the molds and tools required in traditional manufacturing techniques. In other words, 3D printing is likely to end up being another form of factory automation. Manufacturing robots and industrial printers will work in unison—and increasingly without the involvement of workers.

Three-dimensional printers can be used with virtually any type of material, and the technology is finding many important uses outside of manufacturing. Perhaps the most exotic application is in printing human organs. San Diego–based Organovo, a company that specializes in bio-printing, has already fabricated experimental human liver and bone tissue by 3D-printing material containing human cells. The company hopes to produce a complete printed liver by the end of 2014. These initial efforts would produce organs for research or drug testing. Organs suitable for transplant likely remain at least a decade in the future, but if the technology arrives, the implications would be staggering for the roughly 120,000 people awaiting organ transplants in the United States alone.[4] Aside from addressing the shortage, 3D printing would also allow organs to be fabricated from a patient's own stem cells, essentially eliminating the danger of rejection after a transplant.

Food printing is another popular application. Hod Lipson suggests in his 2013 book *Fabricated: The New World of 3D Printing* that digital cuisine may turn out to be 3D printing's "killer app"—in other words, the application that motivates huge numbers of people to go out and buy a home printer.[5] Food printers are currently used to produce designer cookies, pastries, and chocolates, but they also have the potential to combine ingredients in unique ways, synthesizing unprecedented tastes and textures. Perhaps someday 3D food printers will be ubiquitous in home and restaurant kitchens, and gourmet chefs will be subjected to the same type of winner-take-all digital market that professional musicians currently face.

The biggest disruption of all could come when 3D printers are scaled up to construction size. Behrokh Khoshnevis, an engineering professor at the University of Southern California, is building a massive 3D printer capable of fabricating a house in just twenty-four hours. The machine runs on temporary rails alongside the construction site and has a huge printer nozzle that deposits layers of concrete under computer control. The process is entirely automated, and the resulting walls are substantially stronger than those built using traditional techniques.[6] The printer could be used to build homes, office buildings, and even multi-level towers. Currently the machine builds only the structure's concrete walls, leaving workers to install doors, windows, and other fittings. However, it is easy to imagine future construction printers being upgraded to handle multiple materials.

The impact of 3D printing on manufacturing may be relatively muted simply because factories are already highly automated. The story could be very different in the construction industry. Building wood-frame homes is one of the most labor-intensive areas of the economy and offers one of the few remaining occupational opportunities for relatively unskilled workers. In the United States alone nearly 6 million people are employed in the construction sector, while the International Labour Organisation estimates that global construction employment is nearly 110 million.[7] Three-dimensional construction printers might someday result in better and cheaper homes, as well as radically new architectural possibilities—but the technology could also eliminate untold millions of jobs.

Autonomous Cars

The self-driving car entered the final stretch on the road that would take it from science fiction to everyday reality on March 13, 2004. That date marked the first DARPA Grand Challenge—a race that the Defense Advanced Research Projects Agency hoped would help jump-start progress in the development of autonomous military vehicles. Fifteen robotic vehicles set off on a course that began near the town of Barstow, California, and wound its way 150 miles across the Mojave Desert. At stake was a $1 million prize for the first contestant to cross the finish line. The results were underwhelming.

None of the vehicles managed to complete even 10 percent of the course. The best effort came from Carnegie Mellon University's modified Humvee, which careened off the road after just seven and a half miles and plunged into an embankment. DARPA declared the race a bust and kept its money.

The agency saw promise, however; it scheduled a rematch and upgraded the prize to $2 million. The second race was held on October 8, 2005, and required the robotic vehicles to navigate more than one hundred sharp turns, pass through three tunnels, and trek across a mountain pass with sheer drop-offs on both sides of the winding dirt path. The progress was astonishing. After just eighteen months of continued development, five of these vehicles leapt literally from the ditch to the finish line. The winning entry, a modified Volkswagen Touareg designed by a team led by Stanford University's Sebastian Thrun, completed the race in just under seven hours. Carnegie Mellon's refined Humvee design crossed the finish line about ten minutes later. Two other vehicles followed within half an hour.

DARPA staged yet another challenge in November 2007. This time the agency created an urban setting in which robotic vehicles shared the road with a fleet of thirty Ford Tauruses manned by professional drivers. The self-driving cars had to obey traffic regulations, merge into traffic, park, and negotiate busy intersections. Six out of thirty-five robotic vehicles managed to complete the course. Stanford's car was once again first over the finish line but was later demoted to second place after judges analyzed the data and subtracted points for infractions of California's driving laws.[8]

Google's autonomous-car project got its start in 2008. Sebastian Thrun, who had come to the company a year earlier to work on the Street View project, was put in charge, and Google began to rapidly hoover up the best engineers who had worked on the vehicles entered in the DARPA races. Over the course of two years, the team developed a modified Toyota Prius packed with sophisticated equipment, including cameras, four separate radar systems, and an $80,000 laser range finder capable of creating a complete three-dimensional model of the car's environment. The cars can track vehicles, objects, and pedestrians; read traffic signs; and handle nearly any driving scenario. As of 2012, Google's autonomous fleet had driven over 300,000 accident-free miles on roads ranging from freeways jammed with stop-and-go traffic to San Francisco's famously convoluted Lombard Street. In October 2013, the company released data showing that its cars consistently outperformed the typical human driver in terms of smooth acceleration and braking, as well as general defensive driving practices.[9]

Google's project has had a galvanizing effect on the automotive industry. Virtually every major car manufacturer has since announced plans to implement at least a semi-autonomous driving system within the next decade or so. The current leader is Mercedes-Benz. The 2014 S-Class is already capable of driving autonomously in stop-and-go city traffic or on the Autobahn at up to 120 miles per hour. The system locks onto either lane markings or the car ahead and handles steering, acceleration,and braking. Mercedes has initially chosen to take a cautious approach, however, and the driver is required to keep his or her hands on the steering wheel at all times.

Indeed, the systems under development within the automotive industry are almost universally geared toward partial automation—the idea being that the human driver always maintains ultimate control. Liability in the event of an accident may be one of thorniest potential issues surrounding fully automated cars; some analysts have suggested that there might be ambiguity as to who would be responsible. Chris Urmson, one of the engineers who led Google's car project, said at an industry conference in 2013 that such concerns are misplaced, and that current US law makes it clear that the car's manufacturer would be responsible in the event of an accident. It's hard to imagine anything the automotive industry would fear more. Deep-pocketed manufacturers would make irresistible targets for attorneys wielding product liability claims. Urmson went on to argue, however, that because automated cars continuously collect and store operational data that would offer a comprehensive picture of the car's environment up to the moment of the accident, it would be nearly impossible to succeed with a frivolous lawsuit.[10] Still, no technology is 100 percent reliable, and it's therefore inevitable that an autonomous system will eventually cause an accident that confronts its manufacturer with a daunting liability judgment. One possible solution would be laws placing reasonable limits on such lawsuits.

The semi-autonomous approach creates problems of its own, however. None of the systems are yet capable of handling every situation. Google's corporate blog noted in 2012 that, while progress on self-driving cars has been encouraging, "there's still a long road ahead" and that its cars still "need to master snow-covered roadways, interpret temporary construction signals and handle other tricky situations that many drivers encounter."[11] The grey area where a car may need to detect that it is encountering an unmanageable situation and then successfully return control to the driver probably represents the technology's greatest weakness. The engineers working on the systems have found that it takes about ten seconds to alert the driver and ensure that he or she regains control of the vehicle. In other words, the system has to anticipate a potential problem well before the car actually gets into trouble; accomplishing that with a high degree of reliability is a substantial technical challenge. This would be made worse if drivers were not required to keep their hands on the wheel during automated driving. One Audi official noted that when the system being developed by the company is engaged, the driver is "not allowed to sleep, read a newspaper, or a use a laptop."[12] It's unclear how the company plans to enforce that—or if using a smart phone, watching a movie, or engaging in any number of other distractions would be allowed.

Once such hurdles are overcome, autonomous cars offer enormous potential, especially in terms of improved safety. In 2009, there were about 11 million automobile accidents in the United States, and about 34,000 people were killed in collisions. Globally, about one and a quarter million people are killed on roads each year.[13] The National Transportation Safety Board estimates that 90 percent of accidents occur primarily because of human error. In other words, an enormous number of lives might be saved by truly reliable self-driving technology. Preliminary data suggests that the collision avoidance systems now available in some cars are already having a positive impact. A study of insurance claim data by the Highway Loss Data Institute found that some Volvo models equipped with such systems experienced roughly 15 percent fewer accidents than comparable cars without the technology.[14]

Aside from accident avoidance, self-driving car proponents point to many other potential upsides. Autonomous cars will be able to communicate and collaborate with each other. They might travel in convoys, riding in each other's draft to save fuel. High-speed coordination on freeways would reduce, or perhaps even virtually eliminate, traffic jams. Here, I think the hype is running substantially ahead of any near-term reality. Benefits of this type rely heavily on a network effect: a substantial fraction of the cars on the road would need to be autonomous. The obvious reality is that a great many drivers are going to be, at best, ambivalent about self-driving technology. A lot of people simply like to drive. Enthusiast magazines like *Motor Trend* and *Car and Driver* have millions of subscribers. What, after all, is the point of owning "the ultimate driving machine" if you aren't going to drive it? Even among drivers who embrace the technology, adoption is likely to be quite gradual. One consequence of soaring income inequality and decades of stagnant incomes is that new cars are becoming increasingly unaffordable to a large fraction of the population. Indeed, recent data suggests that American consumers are in no rush to trade in the vehicles they have. In 2012, the average car on the road in the United States was nearly eleven years old—an all-time record.

In some cases, a mixture of human and robotic drivers might actually lead to more problems. Think about the last aggressive driver you encountered—the person who cut you off or perhaps weaved recklessly between lanes on the highway. Now imagine that person sharing the road with autonomous cars he or she knows are programmed to be flawlessly defensive in all situations. Such "wolf among sheep" scenarios might invite even more risky behavior.

The most optimistic boosters of self-driving car technology expect a major impact within five to ten years. I suspect that technical challenges, social acceptance, and obstacles related to liability and regulation may make such projections seem overly optimistic. Nonetheless, I think there's little doubt that truly autonomous—or in other words "driverless"—vehicles will eventually arrive. When they do, they will have the potential to revolutionize not just the automotive industry but entire sectors of our economy and job market, as well as the fundamental relationship between people and automobiles.

Perhaps the most important thing to understand about a future in which your car is fully autonomous is that it probably *won't be your car*. Most people who have given serious thought to the optimal role of self-driving cars seem to agree that, at least in densely populated areas, they are likely to be a shared resource. This has been Google's intent from the start. As Google co-founder Sergey Brin explained to the *New Yorker*'s Burkhard Bilger, "[L]ook outside, and walk through parking lots and past multilane roads: the transportation infrastructure dominates. It's a huge tax on the land."[15]

Google hopes to smash the prevailing owner-operator model for the automobile. In the future, you'll simply reach for your smart phone or other connected device and call for a self-driving car whenever you need it. Rather than spending 90 percent or more of their time parked, cars will see much higher utilization rates. That change alone would unleash a real-estate revolution in cities. Vast stretches of space now earmarked for parking would become available for other uses. To be sure, self-driving cars would still need to be stored somewhere when not in use, but there would be no need for random egress; the cars could be packed end-to-end. If you call for a car, and there isn't already one on the road close to your location, you'll simply get the next vehicle in line.

There are, of course, some reasons to be skeptical that urban cars will ultimately evolve into public resources. For one thing, it would be directly at odds with the goals of the automotive industry, which would like each household to own at least one car. For another, in order for this model to work, commuters would have to share the cars at peak times; otherwise they might be so scarce and expensive during busy periods that many people couldn't afford a ride. A related problem is safety in a shared car. Even if the vehicle's software is able to solve the logistics issues and provide efficient and timely service, a small car is, after all, a much more intimate space to share with complete strangers than a bus or train. It's easy to imagine solutions to this problem, however. For example, cars designed to be shared by solo travelers could simply be divided into compartments. You wouldn't even need to see or be aware of others sharing your car. To avoid a feeling of being closed in, virtual windows could be mounted on the dividing walls; high resolution screens would display images captured by cameras mounted on the exterior of the car. By the time self-driving cars are in routine operation, the hardware to accomplish all this will be remarkably inexpensive. The vehicle would stop, a green light would flash on one of the doors, and you would get in and ride to your destination just as if you were traveling alone. You'd be sharing the vehicle, but riding in your own virtual commuter pod. Other vehicles might be designed to carry groups (or more sociable solo travelers), or perhaps the barriers could slide away upon mutual consent.*

* One problem with shared automated cars, especially if they had private compartments, would likely be keeping the vehicles clean. This is a common problem on buses and subways, and in the absence of a driver (or other passengers) some people might not be on their best behavior.

Then, again, the commuter pod might not need to be "virtual." In May 2014, Google announced that the next phase of its research into self-driving cars would focus on the development of two-passenger electric vehicles with a top speed of 25 miles per hour and specifically geared toward urban environments. Passengers would call for the car and set its destination with a smart phone app. Google engineers have come to the conclusion that returning the vehicle to the driver's control in the event of an emergency is unfeasible, and the vehicles will be fully automated—with no steering wheel or brake pedal. In an interview with John Markoff of the *New York Times,* Sergey Brin highlighted the company's dramatic departure from the more "incremental" designs being pursued by the major auto manufacturers, saying "that stuff seems not entirely in keeping with our mission of being transformative."[16]

The market might also create other solutions geared toward sharing automated vehicles. Kevin Drum of *Mother Jones,* who thinks that "genuine self-driving cars will be available within a decade and that they'll be big game changers,"[17] has suggested that it might be possible to purchase a share in a car service, with guaranteed availability, for a fraction of what it would cost to buy a vehicle. In other words, you would share the car only with fellow subscribers to a service, rather than with the public at large.*

If the sharing model does prevail, higher utilization for each car would, of course, mean fewer vehicles relative to the population. Environmentalists and urban planners would likely be overjoyed;

---

* If the sharing model doesn't take hold, then automated cars could actually have a negative impact in congested areas. If you own an automated car and need to visit an area where parking is scarce and expensive, you might choose to have the car simply circle around and then pick you up once you complete your business. Or perhaps you might send it to cool its heels in an adjacent residential neighborhood rather than pay for parking. You might even have downloaded an illicit software application that allows your car to park illegally and then zip away in the nick of time if it detects the approach of an official-looking vehicle.

automobile manufacturers not so much. Beyond the prospect of fewer cars per capita, there could also be a significant threat to luxury automotive brands. If you don't own the car and will use it for only a single trip, you have little reason to care what make or model it is. Cars could cease to be status items, and the automobile market might well become commoditized. For these reasons, I think it's a good bet that the auto manufacturers will cling pretty tightly to keeping someone in the driver's seat—even if he or she rarely touches the controls. Automotive manufacturers could be poised to face the kind of dilemma that powerful companies often encounter when disruptive technologies come along. The company is forced to choose between protecting the business that provides revenue today and in the near future—or helping to propel an emerging technology that may ultimately devalue or even destroy that legacy business. History shows that companies nearly always choose to protect their established revenue streams.* If the kind of revolution that Brin envisions is to unfold, it may have to arise outside the automotive industry. And, of course, Brin may be in exactly the right place to make that happen.

If the individual-ownership model for cars ultimately falls, the impact on broad swathes of the economy and job market would be extraordinary. Think of all the car dealers, independent repair shops, and gas stations within a few miles of your home. Their existence is all tied directly to the fact that automobile ownership is widely distributed. In the world that Google envisions, robotic cars will be concentrated into fleets. Maintenance, repair, insurance, and fueling would likewise be centralized. Untold thousands of small businesses, and the jobs associated with them, would evaporate. To get a sense of just how many jobs might be at risk, consider that, *in Los Angeles alone,* about 10,000 people work in car washes.[18]

* Microsoft clinging to its massive Windows-based revenue stream and failing to get a toehold in the smart phone and tablet markets is a classic example of this.

The most immediate employment impact, of course, would be on those who drive for a living. Taxi driving jobs would evaporate. Bus driving might be automated, or perhaps buses will simply disappear, replaced by a better and more personalized form of public transportation. Delivery jobs might also disappear. Amazon, for example, is already experimenting with same-day deliveries to lockers in fixed locations. Why not put the lockers on wheels? An automated delivery van might send a text message to the customer a few minutes before its arrival and then simply wait for the customer to enter a code and retrieve the package.*

Indeed, I think that commercial fleets could be one of the first places where we see widespread adoption of automated vehicles. The companies that own and operate these fleets already face enormous liability. A single mistake on the part of a single driver can make for a very bad day. Once the technology has a solid track record and the data demonstrates a clear safety and reliability advantage, there will be a very powerful incentive to automate these vehicles. In other words, the first place where self-driving cars make serious inroads might be exactly the area that directly impacts the most jobs.

I've seen many suggestions that heavy, long-haul trucks might also be fully automated in the relatively near future. Here, again, I think progress is likely to be far more measured. While the trucks may indeed soon be able to essentially drive themselves, the staggering destructive potential of these vehicles probably means that someone is going to remain in the driver's seat for the foreseeable future. Experiments with automated convoys, where a truck is programmed to follow the vehicle ahead of it, have already been

---

* This strikes me as far more viable than the drone-based delivery idea that Amazon unveiled in a 2013 episode of CBS's *60 Minutes*. No technology can be made 100 percent reliable. Amazon's business is so vast that, in order to have a meaningful impact, an enormous number of drone-based deliveries would have to occur. Even a very small error rate multiplied by the huge number of flights would likely result in a continuous stream of unfortunate incidents. An incident involving a five-pound payload potentially suspended hundreds of feet in the air is not an incident you would want to have.

successful and may have an important role in the military or in less populated areas. In a 2013 interview with *Time* magazine's David Von Drehle, one trucking company executive made the important point that the United States' crumbling infrastructure presents a significant obstacle to making full automation viable.[19] Truck drivers have to routinely deal with the reality that our roads and bridges are basically falling apart, and that they are constantly being patched up. As I suggested in Chapter 1, getting rid of truck drivers entirely might also make deliveries of food and other critical supplies susceptible to hacking or cyber attack.

Excepting perhaps electricity, there is no other single innovation that has been more central to the development of the American middle class—and the established fabric of society in nearly all developed countries—than the automobile. The true driverless vehicle has the potential to completely upend the way we think about and interact with cars. It could also vaporize millions of solid middle-class jobs and destroy untold thousands of businesses. A small preview of the conflict and social upheaval that are sure to accompany the rise of self-driving cars can be found in the conflagration surrounding Uber, a start-up company that allows people to call for a ride using their smart phone. The company has been embroiled in controversy and litigation in nearly every market it has entered. In February 2014, Chicago taxicab operators filed a lawsuit against the city, claiming that Uber is devaluing nearly 7,000 city-issued operating licenses with a total market value of over $2.3 billion.[20] Imagine the uproar when Uber's cars start arriving without drivers.

As jobs evaporate and median incomes stagnate—or perhaps even fall—we run the risk that a large and growing fraction of our population will no longer have sufficient discretionary income to continue propelling vibrant demand for the products and services that the economy produces. In the next chapter we'll examine this danger, and see how it might ultimately threaten economic growth, and perhaps even precipitate a new crisis.

Chapter 8

CONSUMERS, LIMITS TO GROWTH . . . AND CRISIS?

There is an often-told story about Henry Ford II and Walter Reuther, the legendary head of the United Auto Workers union, jointly touring a recently automated car manufacturing plant. The Ford Motor Company CEO taunts Reuther by asking, "Walter, how are you going to get these robots to pay union dues?" Reuther comes right back at Ford, asking, "Henry, how are you going to get them to buy your cars?"

While that conversation probably never actually took place, the anecdote nonetheless captures a key concern about the ultimate impact of widespread automation: workers are also consumers, and they rely on their wages to purchase the products and services produced by the economy. Perhaps more than any other economic sector, the automotive industry has showcased the importance of this dual role. When the original Henry Ford ramped up production of the Model T in 1914, he famously doubled wages to $5 per day—and, in so doing, ensured that his workers would be able to afford to buy the cars they were building. From that genesis, the rise of the automotive industry would go on to become inextricably intertwined with the creation of a massive American middle class. As we saw in Chapter 2, there is evidence to suggest that this powerful symbiosis between rising incomes and robust, broad-based consumer demand is now in the process of unwinding.

A Thought Experiment

To visualize the most extreme possible implications of Reuther's warning, consider a thought experiment. Imagine that Earth is suddenly invaded by a strange extraterrestrial species. As thousands of the creatures stream off their massive spacecraft, humanity comes to understand that the visitors have not come to conquer us, or to extract our resources, or even to meet our leader. The aliens, it turns out, have come to work.

The species has evolved along a path dramatically different from that of human beings. The alien society is roughly comparable to that of social insects, and the creatures aboard the spacecraft are drawn entirely from the worker caste. Each individual is highly intelligent and capable of learning language, solving problems, and even exhibiting creativity. However, the aliens are driven by a single—and overwhelming—biological imperative: fulfillment comes only from performing useful work.

The aliens have no interest in leisure, entertainment, or general intellectual pursuits. They have no concept of a home or personal space, private property, money or wealth. If they need to sleep they do so standing in their workplaces. They are indifferent even to the food they eat, as they have no sense of taste. The aliens reproduce asexually and reach full maturity within months. They have no need to attract mates and no desire to stand out as individuals. The aliens serve the colony. They are driven to work.

Gradually, the aliens integrate into our society and economy. They are eager to work, and they demand no wages. Work, for the aliens, is its own reward; indeed, it is the only reward they can conceive of. The sole cost associated with their employment is the provision of some type of food and water—and given this, they begin to reproduce rapidly. Businesses of all sizes quickly begin to deploy the extraterrestrials in a variety of roles. They start off in more routine, low-level jobs but rapidly demonstrate the ability to take on more complex work. Gradually, the aliens displace human workers. Even those business owners who initially resist replacing people with aliens eventually have little choice but to make the transition once their competitors do so.

Among humans, unemployment begins to rise relentlessly while incomes for those who still have jobs stagnate and even begin to fall as competition for jobs increases. Months and then years pass, and unemployment benefits run out. Calls for government intervention result only in gridlock. In the United States, Democrats call for restrictions on employing aliens; Republicans, lobbied heavily by big business, block these initiatives and point out that the aliens have spread across the globe. Any limitations on the ability of American businesses to employ the aliens would put the country at a staggering competitive disadvantage.

The public becomes increasingly fearful about the future. Consumer markets become deeply polarized. A small number of people—those who own a successful business, hold large investments, or have safe executive-level jobs—have been doing extremely well as business profitability has increased. Sales of luxury goods and services are booming. For the rest, it's the dollar store economy. As more people are unemployed, or become fearful that they will soon lose their jobs, frugality becomes tantamount to survival.

Soon, however, it becomes evident that those dramatic increases in business earnings are unsustainable. The profits have come almost entirely from cutting labor costs. Revenues are flat, and soon they begin to fall. The aliens, of course, buy nothing. Human consumers increasingly turn away from any purchase that is not absolutely essential. Many businesses that produce nonessential goods and services eventually begin to fail. Savings and credit lines are exhausted. Homeowners become unable to pay their mortgages; tenants fail to make rent payments. Default rates for home loans, business loans, consumer debt, and student loans soar. Tax revenues collapse even as demand for social services rises dramatically, threatening the solvency of governments. Indeed, as a new financial crisis looms, even the prosperous elite will cut back on their consumption: rather than expensive handbags or luxury cars, they will soon be more interested in buying gold. The alien invasion, it seems, has not turned out to be so benign after all.

Machines Do Not Consume

The alien invasion parable is admittedly extreme. Perhaps it would work as the plot for a really low-budget science fiction movie. Nonetheless, it captures the theoretical endpoint of a relentless progression toward automation—at least in the absence of policies designed to adapt to the situation (more on that in Chapter 10).

The primary message this book has delivered so far is that accelerating technology is likely to increasingly threaten jobs across industries and at a wide range of skill levels. If such a trend develops, it has important implications for the overall economy. As jobs and incomes are relentlessly automated away, the bulk of consumers may eventually come to lack the income and purchasing power necessary to drive the demand that is critical to sustained economic growth.

Every product and service produced by the economy ultimately gets purchased (consumed) by someone. In economic terms, "demand" means a desire or need for something, backed by the ability and willingness to pay for it. There are only two entities that create final demand for products and services: individual people and governments. Individual consumer spending is typically at least two-thirds of GDP in the United States and roughly 60 percent or more in most other developed countries. The vast majority of individual consumers, of course, rely on employment for nearly all of their income. Jobs are the primary mechanism through which purchasing power is distributed.

To be sure, businesses also purchase things, but that is not final demand. Businesses buy inputs that are used to produce something else. They may also buy things to make investments that will enable future production. However, if there is no demand for what the business is producing, it will shut down and stop buying inputs. A business may sell to another business; but somewhere down the line, that chain has to end at a person (or a government) buying something just because they want it or need it.

The essential point is that a worker is also a consumer (and may support other consumers). These people drive final demand. When a worker is replaced by a machine, that machine does not go out and consume. The machine may use energy and spare parts and require maintenance, but again, those are business inputs, not final demand. If there is no one to buy what the machine is producing, it will ultimately be shut down. An industrial robot in an auto manufacturing plant will not continue running if no one is buying the cars it is assembling.*

So if automation eliminates a substantial fraction of the jobs that consumers rely on, or if wages are driven so low that very few people have significant discretionary income, then it is difficult to see how a modern mass-market economy could continue to thrive. Nearly

---

* Not all robots are used in production, of course. There are also consumer robots. Suppose you someday own a personal robot, capable of doing things around the house. It may "consume" electricity and require repair and maintenance. However, in economic terms, *you* are the consumer—not the robot. You need a job/income or you won't be able to pay for the operating costs of your robot. Robots don't drive final consumption—people do. (Assuming, of course, that robots are not truly intelligent, sentient, and accorded the economic freedom that would be necessary for them to act as consumers. We'll consider that speculative possibility in the next chapter.)

all the major industries that form the backbone of our economy (automobiles, financial services, consumer electronics, telecommunications services, health care, etc.) are geared toward markets consisting of many millions of potential customers. Markets are driven not just by aggregate dollars but also by unit demand. A single very wealthy person may buy a very nice car, or perhaps even a dozen such cars. But he or she is not going to buy thousands of automobiles. The same is true for mobile phones, laptop computers, restaurant meals, cable TV subscriptions, mortgages, toothpaste, dental checkups, or any other consumer good or service you might imagine. In a mass-market economy, the distribution of purchasing power among consumers matters a great deal. Extreme income concentration among a tiny sliver of potential customers will ultimately threaten the viability of the markets that support these industries.

Inequality and Consumer Spending: The Evidence So Far

In 1992, the top 5 percent of US households in terms of income were responsible for about 27 percent of total consumer spending. By 2012, that percentage had risen to 38 percent. Over the same two decades, the share of spending attributed to the bottom 80 percent of American consumers fell from about 47 percent to 39 percent.[1] By 2005, the trend toward increased concentration of both income and spending was so obvious and relentless that a team of stock market analysts at Citigroup famously wrote a series of memos intended only for their wealthiest clients. The analysts argued that the United States was evolving into a "plutonomy"—a top-heavy economic system where growth is driven primarily by a tiny, prosperous elite who consume an ever larger fraction of everything the economy produces. Among other things, the memos advised wealthy investors to shy away from the stocks of companies catering to the rapidly dissolving American middle class and instead focus on purveyors of luxury goods and services aimed at the richest consumers.[2]

The data demonstrating the American economy's decades-long march toward ever-increasing concentration of income is unequivocal, but it contains within it a fundamental paradox. Economists have long understood that the wealthy spend a smaller fraction of their income than the middle class and, especially, the poor. The lowest-income households have little choice but to spend nearly everything they manage to bring in, while the truly rich would likely find it impossible to consume at a similar rate even if they tried. The clear implication is that, as income increasingly concentrates among the wealthy few, we should expect less robust overall consumption. The tiny slice of the population that's hoovering up more and more of the country's total income simply isn't going to be able to spend it all, and that ought to be obvious in the economic data.

The historical reality, however, turns out to be quite different. Over the three and a half decades between 1972 and 2007, average spending as a percentage of disposable income increased from roughly 85 percent to more than 93 percent.[3] For most of that period, consumer spending was not only by far the largest component of American GDP—it was also the fastest growing. In other words, even as income became ever more unequal and concentrated, consumers managed to somehow actually increase their overall spending, and their profligacy was the most important factor powering the growth of the American economy.

In January 2014, economists Barry Cynamon and Steven Fazzari, of the Federal Reserve Bank of St. Louis and Washington University in St. Louis respectively, published research that delved into the paradox of increasing income inequality coupled with rising consumer spending. Their primary conclusion was that the decades-long uptrend in consumer spending was powered largely by increased debt taken on by the lower 95 percent of American consumers. Between 1989 and 2007 the ratio of debt to income for this vast majority roughly doubled from just over 80 percent to a peak of nearly 160 percent. Among the wealthiest 5 percent, the same ratio remained relatively constant at around 60 percent.[4] The steepest increase in debt levels tracked closely with the housing bubble and easy access to home equity credit in the years leading up to the financial crisis.

That relentless borrowing on the part of nearly the entire American population was, of course, ultimately unsustainable. Cynamon and Fazzari argue that "financial fragility created by unprecedented borrowing triggered the Great Recession when the inability to borrow more forced a drop in consumption."[5] As the crisis unfolded, overall consumer expenditures plunged by about 3.4 percent, a collapse in consumption unmatched during any recession since World War II. The spending decline was also especially long-lived; it took nearly three years for overall consumption to return to its pre-crisis level.[6]

Cynamon and Fazzari found a marked difference between the two income groups both during and after the Great Recession. The top 5 percent were able to moderate their spending by drawing on other resources during the recession. The bottom 95 percent were essentially tapped out and had little choice but to cut back dramatically. The economists also discovered that the subsequent recovery in consumer spending has been powered entirely by the top of the income distribution. By 2012, the top 5 percent had increased their spending by about 17 percent, after adjusting for inflation. The bottom 95 percent had seen no recovery at all; consumption remained mired at 2008 levels. Cynamon and Fazzari see few prospects for a meaningful recovery among the majority of consumers and "fear that the demand drag from rising inequality that was postponed for decades by bottom 95 percent borrowing is now slowing consumption growth and will continue to do so in coming years."[7]

In corporate America it has become increasingly evident that, when it comes to domestic customers, all the action is at the top. In virtually every industry sector that caters directly to American consumers—from home appliances to restaurants and hotels to retail stores—the mid-range is struggling with stagnant or declining sales, while companies that target top-tier consumers continue to thrive. Some business leaders are beginning to recognize the obvious threat to mass-market products and services. In August 2013, John Skipper, the president of ESPN, the cable and satellite sports network that ranks as the world's most valuable media brand, said that income stagnation represented the greatest single threat to the future of his company. The cost of cable TV service in the United States has soared by about 300 percent over the past fifteen years, even as incomes have remained flat. Skipper noted that "ESPN is a mass-product," and yet the service could eventually be out of reach for a large fraction of its audience.[8]

As America's largest retailer, Walmart has become a bellwether for the middle- and working-class consumers who flock to its stores in search of low prices. In February 2014, the company released an annual sales forecast that disappointed investors and caused the stock to drop sharply. Sales at established stores (those open for at least a year) had fallen for the fourth quarter in a row. The company warned that cuts to the US food stamp program (officially known as the Supplemental Nutrition Assistance Program) as well as increases in payroll taxes were poised to hit hard at low-income shoppers. About one in five Walmart customers rely on food stamps, and evidence suggests that many of these people are stretched to the point where they have virtually no discretionary income.

In the wake of the Great Recession, Walmart stores routinely see an explosion of activity just after midnight on the first of each month—the day that electronic benefits transfer (EBT) cards are reloaded by the government. By the end of the month, Walmart's lowest-income customers have quite literally run out of food and other essentials, so they load up their shopping carts and line up in anticipation of a credit from the food stamp program that generally comes through shortly after midnight.[9] Walmart has also suffered from increased competition from dollar stores; in many cases its customers are turning to these outlets not necessarily because overall prices are lower but, rather, because the stores offer smaller quantities that help them stretch their few remaining dollars as they struggle to make it through the final days of the month.

Indeed, throughout the private sector, the recovery has largely been characterized by soaring corporate earnings coupled with often underwhelming revenues. Corporations have achieved dizzying levels of profitability, but they have accomplished this primarily by cutting labor costs—not by selling more of the goods and services they produce. This shouldn't come as a surprise: take a moment to look back at Figures 2.3 and 2.4 in Chapter 2. Corporate profits as a share of GDP reached unprecedented heights even as labor's share of national income plunged to a record low. To me, this suggests that a great many American consumers are struggling to purchase the products and services that companies are producing. Figure 8.1, which shows how general US corporate earnings recovered rapidly and have been pulling away from retail sales over the course of the recovery, makes the story still more clear.* Keep in mind that, as we saw previously, the gradual spending recovery has been powered entirely by consumers in the top 5 percent of the income distribution.

The Wisdom of the Economists

Despite the evidence suggesting that a huge percentage of American consumers simply don't have sufficient income to create adequate demand for the products and services produced by the economy, there is no general agreement among economists that income inequality is creating a substantial drag on economic growth. Even among America's leading progressive economists—nearly all of whom would likely agree that a lack of demand is a primary problem

* It's important to note that retail sales are only a small fraction of overall consumption, or what is technically called personal consumption expenditure (PCE). PCE is usually around 70 percent of US GPD and includes all the products and services that consumers purchase, as well as housing expenditures—either rent or "imputed rent" (a measure used for owner-occupied dwellings).

Figure 8.1. US Corporate Profits Versus Retail Sales During Recovery from the Great Recession

SOURCE: Federal Reserve Bank of St. Louis (FRED).[10]

facing the economy—there is no consensus about the direct impact of inequality.

The Nobel laureate economist Joseph Stiglitz has been perhaps the most vocal proponent of the idea that inequality undermines economic growth, writing in a January 2013 *New York Times* op-ed that "inequality is squelching our recovery" because "our middle class is too weak to support the consumer spending that has historically driven our economic growth."[11] Robert Solow—who won the Nobel Prize in 1987 for his work on the importance of technological innovation to long-term economic growth—seems to largely agree, saying in a January 2014 interview that "increasing inequality tends to hollow out the income distribution, and we lose the solid middle class jobs and steady middle class incomes which provide a reliable flow of consumer demand that keeps industry going and innovating."[12] Paul Krugman, yet another Nobel laureate—and the one with the highest profile as a columnist and blogger for the *New York Times*—disagrees, however, writing in his blog that he wishes he "could sign on to this thesis," but that the evidence doesn't support it.[13*]

Among more conservative economists, the idea that inequality is a significant drag on growth is likely to be dismissed entirely. Indeed, many right-leaning economists are reluctant even to accept the argument that a lack of demand has been the primary problem facing the economy. Instead, throughout the course of the recovery, they have pointed to uncertainty surrounding issues like public debt levels, potential tax increases, increased regulation, or the implementation of the Affordable Care Act. Cutting government spending and reducing taxes and regulation, they say, will spur investor and business confidence, leading to increased investment, economic growth, and employment. This idea—which seems to me to be remarkably divorced from the obvious reality—has been repeatedly disparaged by Krugman as a belief in "the confidence fairy."[14]

My key point here is that professional economists—all of whom have access to the same objective data—are completely unable to agree on what I would characterize as an extraordinarily

---

* Krugman's primary objection relates to the fact that consumers at various points on the income distribution aren't necessarily at that level all the time. Some people might be having an especially good or bad year, and their spending will be more a function of their long-term expectations than their current situation. (This, as we will see shortly, relates to what's called the "permanent income hypothesis.") As a result, Krugman says, looking at the data at any moment in time "tells you nothing at all about what will happen." Krugman points out that "economics is not a morality play," and goes so far as to suggest that we can have "full employment based on purchases of yachts, luxury cars, and the services of personal trainers and celebrity chefs." I am skeptical of this (but see the section on "techno-feudalism" later in this chapter). As I pointed out previously, nearly all the major industries that constitute the modern economy produce mass-market products and services. Yachts and Ferraris just aren't important enough to sustainably offset a broad-based reduction in demand for all the stuff that 99 percent of consumers buy. In any case, production of yachts and Ferraris will increasingly be automated. And how many personal trainers and celebrity chefs do the .01 percent really need?

fundamental economic question: Is a demand shortfall holding back economic growth, and if so, is income inequality an important contributor to the problem? I suspect that the lack of consensus on this question offers a pretty good preview of what we can expect from the economics profession as the technological disruption I've been describing in these pages unfolds. While it's certainly possible that two "scientists" may look at the same data and interpret it differently, in the field of economics the opinions all too often break cleanly along predefined political lines. Knowing the ideological predisposition of a particular economist is often a better predictor of what that individual is likely to say than anything contained in the data under examination. In other words, if you're waiting for the economists to deliver some sort of definitive verdict on the impact that advancing technology is having on the economy, you may have a very long wait.

Beyond the ideological divide in economics, yet another potential problem is the extreme quantification of the field. In the decades since World War II, economics has become extraordinarily mathematical and data-driven. While this certainly has many positive aspects, it is important to keep in mind that there is obviously no economic data streaming in from the future. Any quantitative, data-driven analysis necessarily depends entirely on information gathered in the past, and in some cases, that data may have been collected years or even decades ago. Economists have used all that past data to construct elaborate mathematical models, but most of these trace their origin to the economy of the twentieth century. The limitations of the economists' models were made evident by the near-total failure of the profession to anticipate the 2008 global financial crisis. In a 2009 article entitled "How Did Economists Get It So Wrong?" Paul Krugman wrote that "this predictive failure was the least of the field's problems. More important was the profession's blindness to the very possibility of catastrophic failures in a market economy."[15]

I think there are good reasons to be concerned about a similar failure of the economists' mathematical models as the exponential advance of information technology increasingly disrupts the economy. Adding to the problem is that many of these models employ simplistic—and in some cases seemingly absurd—assumptions about the way consumers, workers, and businesses behave and interact. John Maynard Keynes may have said it best, writing nearly eighty years ago in *The General Theory of Employment, Interest and Money,* the book that arguably founded economics as a modern field of study: "Too large a proportion of recent 'mathematical' economics are merely concoctions, as imprecise as the initial assumptions they rest on, which allow the author to lose sight of the complexities and interdependencies of the real world in a maze of pretentious and unhelpful symbols."[16]

Complexity, Feedback Effects, Consumer Behavior, and "Where Is That Soaring Productivity?"

The economy is an enormously complex system, ripe with a myriad of interdependencies and feedback loops. Change one variable and a variety of effects are likely to cascade through the system, some of which may act to mitigate or counteract the initial change.

Indeed, this propensity for the economy to self-moderate through feedback effects is likely one important reason that the role advancing technology has played in creating inequality remains subject to debate. Economists who are skeptical about the impact of technology and automation often point to the fact that the rise of the robots is not obvious in the productivity data, especially over the short term. For example, in the final quarter of 2013, productivity in the United States fell to an annualized rate of just 1.8 percent, as compared to a much more impressive 3.5 percent in the third quarter.[17] Recall that productivity is measured by dividing the economy's output by the number of hours worked. So if machines and software were indeed substituting for human labor at a rapid clip, you would expect the number of hours worked to fall precipitously—and productivity, in turn, to soar.

The problem with this assumption is that in the real economy, things are not so simple. Productivity does not measure how much a business *could* produce per hour; it measures how much a business actually does produce. In other words, productivity is directly influenced by demand. Output, after all, makes up the numerator of the productivity formula. This is especially important when you consider that most of the economy in developed countries is now made up of service businesses. While a manufacturing company, faced with slack demand, might conceivably choose to keep cranking out products and letting them pile up in inventory or in distribution channels, a service business cannot do this. Within the service sector, output responds immediately to demand, and any business that experiences weak growth in demand for its output is likely to also experience less than impressive productivity growth, unless it immediately cuts its workforce or reduces worker hours sufficiently to keep the numbers in line.

Imagine you own a small business that provides some type of analytic service to large corporations. You have ten employees who are fully engaged. Suddenly, a powerful new software application appears that will allow just eight workers to do the work formerly performed by ten. So you purchase the new software and eliminate two jobs. The robot revolution is at hand! Productivity is poised to soar. But, wait. Now your most important client forecasts a downturn in demand for its own product or service. The contract you were supposed to sign this week never materializes. The near-term future looks grim. You just had a layoff, so you don't want to demoralize your workforce by immediately cutting still more jobs. Before you know it, your eight remaining employees are spending a big chunk of their time watching YouTube videos on your dime. Productivity is tanking!

In fact, this was normally what happened during most past downturns in the United States. Recessions typically saw declining productivity because output fell more than hours worked. However, during the Great Recession of 2007–2009, the opposite happened:

productivity actually increased. Output fell substantially, but hours worked fell even more as businesses very aggressively slashed their workforces, increasing the burden on the remaining workers. The workers who kept their jobs (who certainly feared more cuts in the future) probably worked harder and reduced any time they spent on activities not directly related to their work; the result was an increase in productivity.

In the real economy, of course, scenarios like this play out in countless organizations of all sizes. Somewhere, a firm may be incorporating new technology that increases productivity. Elsewhere another firm may be cutting output in response to slack demand. Averaged together, they result in only a middling overall productivity number. The point is that short-term economic numbers like productivity are likely to be variable and somewhat chaotic. Over the long run, however, the trend will be far more clear. Indeed, we saw evidence for this in Chapter 2; recall that productivity has significantly outpaced wages since the early 1970s.

The impact of weak consumer demand on productivity is just one example of the kind of feedback effect that operates in the economy. There are many others, and they can act in both directions. For example, less than robust consumer demand can also slow the development and adoption of new technology. When businesses make investment decisions, they factor in both the current and the anticipated economic environment. When the outlook is poor or when profits decline, investment in research and development or in new capital expenditures is also likely to fall. The result is that technological progress in subsequent years may be slower than it otherwise would have been.

Another example involves the relationship between labor-saving technology and the wages of relatively unskilled workers. If advancing technology (or some other factor) causes wages to stagnate or even fall, then from management's perspective labor will—at least for a time—become more attractive relative to machines. Consider the fast food industry. In Chapter 1, I speculated that this sector may soon be ripe for disruption as advanced robotic technology is introduced. But this suggests a basic question: Why hasn't the industry already incorporated more automation? After all, putting together hamburgers and tacos hardly seems to be on the forefront of precision manufacturing. The answer, at least in part, is that technology has indeed already had a dramatic impact. While machines have not yet completely substituted for fast food workers on a large scale, technology has deskilled the jobs and made the workers largely interchangeable. Fast food workers are integrated into a mechanized assembly-line process with little training required.* This is why the industry is able to tolerate high turnover rates and workers with minimal skill levels. The effect has been to keep these jobs firmly anchored in the minimum-wage category. And in the United States, after adjusting for inflation, the minimum wage has actually fallen more than 12 percent since the late 1960s.[18]

---

* This "fast food effect" may loom large for skilled workers in many other fields. Long before robots are able to completely replace these workers, technology may deskill the jobs and drive wages down. A classic example of deskilling involves London taxi drivers. Entering this profession requires memorizing an extraordinary amount of information about London's street layout. This is referred to as "The Knowledge" and has been required of cab drivers since 1865. Neuroscientist Eleanor Maguire of University College London found that all this memorization actually resulted in changes to the drivers' brains: London cabbies, on average, developed a larger memory center (or hippocampus) than people in other occupations. The advent of GPS-based satellite navigation has, of course, greatly reduced the value of all that knowledge. Taxi drivers possessing The Knowledge—who drive the famous "black" cabs (no longer black, but now covered in colorful advertising)—still dominate in London, but this is largely due to regulation. Drivers without The Knowledge have to be pre-booked; they are not allowed to be flagged down on the street. Of course, new services like Uber, which lets you book a cab with your smart phone, may soon make the act of flagging down a taxi itself obsolete. The taxi drivers may eventually be replaced completely by automated cars, but long before that happens, technology might well deskill their jobs and lower their wages. Perhaps regulation will save the London cabbies from this fate, but workers in many other fields will not be so lucky.

In his 2001 book *Fast Food Nation,* Eric Schlosser relates how McDonald's was already experimenting with more advanced labor-saving technology in the 1990s. At test sites in Colorado Springs, "robotic drink machines selected paper cups, filled them with ice, and then filled them with soda," while French-fry cooking was fully automated and "advanced computer software essentially ran the kitchen."[19] That all these innovations were not eventually scaled across McDonald's restaurants everywhere may well have something to do with the fact that wages have remained very low. This situation cannot be expected to persist indefinitely, however. Eventually, technology will advance to the point where low wages no longer outweigh the benefits of further automation. Introducing more machines might also convey important benefits beyond simply reducing labor costs, such as improved quality or consistency or the consumer perception that automated preparation is more hygienic. As well, there might be synergies between robotic production and other emerging technologies. For example, today it's easy to imagine a mobile app that allows customers to design a completely custom meal, pay for it in advance, and then expect it to be ready for pickup at a precise time; that would have been fantasy in the 1990s. The upshot of all this is that labor-saving technology in an industry like fast food is unlikely to advance in a consistent, predictable way. Instead, it may remain relatively stable for long periods and then leap forward rapidly once things reach a tipping point that forces a reevaluation of the worker-machine trade-off.

Still another consideration involves the behavior of consumers when they are faced with unemployment or reduced incomes. A change in income that consumers expect to be long-term or permanent will have a much bigger impact on their spending behavior than a short-term one. Economists have an impressive name for this idea—"the permanent income hypothesis"—and it was formalized by Nobel laureate Milton Friedman. For the most part, however, it amounts to simple common sense. If you win a thousand dollars in the lottery, you might spend some of it and save the rest, but you're unlikely to make a major, ongoing change to your spending behavior. After all, it's only a one-time bump in your income. On the other hand, if you get a thousand-dollar raise per month, you might well lease a new car, start eating out more often, or even move to a more expensive home.

Historically, unemployment has been viewed as a short-term phenomenon. If you lose your job but feel confident of finding a new position at comparable pay within a short timeframe, you might choose to simply draw on your savings or use your credit card to continue spending at nearly the same level. During the postwar period, it was common for companies to lay off workers for a few weeks or months and then hire them back as soon as the outlook improved. The situation is obviously now quite different. In the wake of the 2008 financial crisis, the long-term unemployment rate soared to unprecedented levels, and it continues to be very high by historical standards. Even those experienced workers who manage to find a new job very often have to accept a lower-paying position. These realities are not lost on consumers. Accordingly, it seems reasonable to speculate that the perception of what it means to be unemployed may gradually be changing. As more people come to see unemployment as a longer-term—or in some cases perhaps even permanent—situation, this seems likely to amplify the impact of a job loss on their spending behavior. In other words, the historical record is not necessarily a good predictor of the future: as the implications of advancing technology become evident to consumers, they may choose to cut spending more aggressively than has been the case in the past.

The complexity that operates in the real-world economy is, in many ways, somewhat analogous to that of the climate system, which is likewise characterized by a nearly impenetrable web of interdependencies and feedback effects. Climate scientists tell us that, as the amount of carbon dioxide in the atmosphere increases, we should not expect a steady, consistent rise in temperatures. Instead, average temperatures will advance chaotically in an uptrend punctuated by plateaus and, quite possibly, years or even longer periods that are relatively cool. We can also expect an increase in the number of storms and other extreme weather events. A somewhat similar phenomenon may unfold in the economy as income and wealth become progressively more concentrated and an ever larger fraction of consumers struggle with a dearth of purchasing power. Measures like productivity or the unemployment rate will not advance smoothly, and the likelihood of financial crises may well increase. Climate scientists also worry about tipping points. For example, one risk is that rising temperatures might cause the arctic tundra to melt, releasing huge amounts of sequestered carbon and, in turn, causing warming to accelerate. By a similar token, it's possible that at some future point, rapid technological innovations might shift the expectations of consumers about the likelihood and duration of unemployment, causing them to aggressively cut their spending. If such an event occurred, it's easy to see how that could precipitate a downward economic spiral that would impact even those workers whose jobs are not directly susceptible to the technological disruption.

Is Economic Growth Sustainable as Inequality Soars?

As we've seen, overall consumer spending in the United States has so far continued to grow even as it has become ever more concentrated, with the top 5 percent of households now responsible for nearly 40 percent of total consumption. The real question is whether that trend is likely to be sustainable in the coming years and decades, as information technology continues its relentless acceleration.

While the top 5 percent have relatively high incomes, the vast majority of these people are heavily dependent on jobs. Even within these top-tier households, income is concentrated to a staggering degree; the number of genuinely wealthy households—those that can survive and continue spending entirely on the basis of their accumulated wealth—is far smaller. During the first year of recovery from the Great Recession, 95 percent of income growth went to just the top 1 percent.[20]

The top 5 percent is largely made up of professionals and knowledge workers with at least a college degree. As we saw in Chapter 4, however, many of these skilled occupations are squarely in the crosshairs as technology advances. Software automation may eliminate some jobs entirely. In other cases, the jobs may end up being deskilled, so that wages are driven down. Offshoring and the transition to big data–driven management approaches that often require fewer analysts and middle managers loom as other potential threats for many of these workers. In addition to directly impacting households that are already in the top tier, these same trends will also make it harder for younger workers to eventually move up into positions with comparable income and spending levels.

The bottom line is that the top 5 percent is poised to increasingly look like a microcosm of the entire job market: it is at risk of itself being hollowed out. As technology progresses, the number of American households with sufficient discretionary income and confidence in the future to engage in robust spending could well continue to contract. The risk is further increased by the fact that many of these top-tier households are probably more financially fragile than their incomes might suggest. These consumers tend to be concentrated in high-cost urban areas and, in many cases, probably do not feel especially wealthy. Large numbers of them have climbed into the top 5 percent through assortative mating: they have partnered with another high-earning college graduate. However, housing and education costs are often so high for these families that the loss of either job puts the household at substantial risk. In other words, in a two-income household the likelihood that sudden unemployment will lead to a substantial cut in spending is effectively doubled.

As the top tier comes under increasing pressure from technology, there are few reasons to expect that the prospects for the bottom

95 percent of households will improve significantly. Robotics and self-service technology in the service sector will continue to make inroads, holding down wages and leaving relatively unskilled workers with fewer options. Automated vehicles or construction-scale 3D printers may eventually destroy millions of jobs. Many of these workers may experience downward mobility; some will likely choose to leave the labor force entirely. There is a risk that, over time, more households will end up living on incomes that are very close to the subsistence level; we could well see even more shoppers in midnight lines waiting for their EBT cards to be reloaded so they can feed their families.

In the absence of increasing incomes, the only mechanism that will allow the bottom 95 percent to spend more would be to take on more debt. As Cynamon and Fazzari found, it was borrowing that allowed American consumers to continue driving economic growth over the course of the two decades leading up to the 2008 financial crisis. In the wake of that crisis, however, household balance sheets are weak and credit standards have tightened substantially, so a great many Americans cannot finance further consumer spending. Even if credit again begins to flow to these households, that is necessarily a temporary solution. Increased debt is unsustainable without increased income, and there would be an obvious danger that loan defaults might eventually precipitate a new crisis. In the one area where lower-income Americans still have easy access to credit—student loans—the debt burden has already grown to extraordinary proportions and the resulting payments will decimate the disposable income of college graduates (not to mention those who fail to get a degree) for decades to come.

While the argument I'm making here is theoretical, there is statistical evidence to support the contention that inequality can be harmful to economic growth. In an April 2011 report, economists Andrew G. Berg and Jonathan D. Ostry of the International Monetary Fund studied a variety of advanced and emerging economies and came to the conclusion that income inequality is a vital factor affecting the sustainability of economic growth.[21] Berg and Ostry point out that economies rarely see steady growth that continues for decades. Instead, "periods of rapid growth are punctuated by collapses and sometimes stagnation—the hills, valleys, and plateaus of growth." The thing that sets successful economies apart is the duration of the growth spells. The economists found that higher inequality was strongly correlated with shorter periods of economic growth. Indeed, a 10-percentage-point decrease in inequality was associated with growth spells that lasted 50 percent longer. Writing on the IMF's blog, the economists warned that extreme income inequality in the United States has clear implications for the country's future growth prospects: "Some dismiss inequality and focus instead on overall growth—arguing, in effect, that a rising tide lifts all boats." However, "when a handful of yachts become ocean liners while the rest remain lowly canoes, something is seriously amiss."[22]

Long-Term Risks: Squeezed Consumers, Deflation, Economic Crises, and . . . Maybe Even Techno-Feudalism

After I published my first book on the subject of automation in 2009, several readers wrote to me to point out that I had neglected to focus on an important point: robots might indeed drive down wages or cause unemployment, but more efficient production would also make everything much cheaper. So even if your income fell, you'd still be able to continue consuming since prices for the things you wanted to buy would be lower. This seems to make sense, but there are a few notable caveats.

The most obvious issue is that many people might be unemployed entirely and effectively have zero income. In that situation, low prices don't solve their problem. Additionally, some of the most important components of the average household budget are relatively immune to the impact of technology, at least in the short and medium terms. The cost of land, housing, and insurance, for example, are tied to general asset values, which are in turn dependent on the overall standard of living. This is the reason that developing countries like Thailand don't allow foreigners to buy land; doing so might result in prices being bid up to the point where housing would become unaffordable for the country's citizens. As we saw in Chapter 6, health care costs also probably represent a challenge for the robots in the near term. Automation is likely to have the greatest immediate impact on costs in manufacturing and in some discretionary services, especially information and entertainment. Yet, these things are a relatively small part of most household budgets. The big-ticket items—housing, food, energy, health care, transportation, insurance—are much less likely to see rapid, near-term cost reductions. There's a real danger that households will end up being squeezed between stagnant or falling incomes and major-expense items that continue to rise in cost.

Even if technology does eventually manage to reduce prices across the board, there is a critical problem with this scenario. The historical path to prosperity has generally been one of wages increasing faster than prices. If someone from the year 1900 were to travel forward in time and visit a contemporary supermarket, he or she would, of course, be shocked by the high prices. Nonetheless, we now spend a significantly smaller share of our incomes on food than was the case in 1900. Food has become cheaper in real terms even as nominal prices have increased dramatically. This has happened because incomes have increased even more dramatically.

Now imagine the opposite situation: incomes are falling, but prices are falling even faster. In theory, this would also mean your purchasing power was increasing: you should now be able to buy more stuff. In reality, however, deflation is a very ugly economic scenario. The first problem is that a deflationary cycle is quite hard to break. If you know that prices will be lower in the future, why buy now? Consumers hold back, waiting for even lower prices, and that in turn forces even more price cuts as well as reduced production of goods and services. Another problem is that, in practice, it's often difficult for employers to actually lower wages. Instead, they are more likely to cut workers, so deflation is typically associated with soaring unemployment, and again, that eventually leads to a lot of consumers with no income at all.

The third major problem is that deflation makes debt unmanageable. In a deflationary economy, your income may be falling (assuming you're lucky enough to have an income at all), the value of your house is likely falling, and the stock market may well be falling. Your mortgage, car, and student loan payments, however, are not going to fall. Debts are fixed in nominal terms, so as incomes decline, borrowers get squeezed and have even less discretionary income to spend. Governments likewise run into trouble because tax revenues plunge. If the situation continues, eventually loan defaults are likely to soar and a banking crisis may well loom. Deflation is really not something we should wish for. History suggests that the ideal is a mildly inflationary trajectory where incomes grow faster than consumer prices, making the things we want to buy more affordable over time.

Either of these two scenarios—households squeezed between stagnant incomes and rising costs, or outright deflation—has the potential to eventually unleash a severe recession as consumers cut back on their discretionary spending. As I suggested previously, there is also the risk that the unfolding technological disruption could fundamentally change consumer spending behavior as more and more people come to quite rationally fear the prospect of long-term unemployment or even a forced early retirement. In such an event, the short-term fiscal policies typically adopted by governments to combat an economic downturn, such as increased government spending or one-time rebates to taxpayers, might not be especially effective. These policies are intended to inject immediate demand into the economy in order to "prime the pump" in the hope of initiating a self-reinforcing recovery that will lead to increased employment. However, if new automation technologies allow businesses to meet this increased demand without hiring many workers, then the impact on unemployment might well be disappointing. Monetary action by central banks would suffer from a similar problem: more money might be printed, but in the absence of hiring there would be no mechanism to get more purchasing power into the hands of consumers.* In short, conventional economic policies might do very little to directly address consumers' fears about their long-term income continuity.

There is also the risk of a new banking and financial crisis as households are increasingly unable to make payments on their debts. Even a relatively small percentage of bad loans can put a great deal of stress on the banking system. The 2008 financial crisis was precipitated when borrowers who had taken out subprime loans began to default en masse in 2007. While the number of subprime loans

---

* When a central bank like the Federal Reserve "prints money," it normally purchases government bonds. When it settles the transaction, it deposits money into the bank account of whomever it bought the bonds from. This is newly created money: it just appears out of nowhere. Once this new money is in the banking system, the idea is that banks can then loan it out. This is what's known as fractional reserve banking. Banks have to keep a small percentage of the new money on hand, but they're allowed to loan out most of it. The way things are supposed to work is that the banks loan the new money to businesses that can then expand and hire more people. Or the banks might loan to consumers who spend the money, thereby creating new demand. Either way, jobs should be created and money (purchasing power) will flow to consumers. Eventually, the money once again gets deposited in a bank and then most of it can yet again be loaned out—and so on. In this way, the newly created money cascades through the economy, multiplying and generally being fruitful. However, if automation technology eventually makes it possible for businesses to expand or meet new demand without significant hiring, or if demand is so weak that businesses aren't interested in borrowing, then little of the newly created money will find its way to consumers, and so it won't get spent and it won't multiply in the intended fashion. It will just slosh around in the banking system. This is more or less what occurred during the 2008 financial crisis—not because of job automation, but because the banks could not find creditworthy borrowers, and/or no one wanted to borrow anyway. Everyone just wanted to hold onto their cash. Economists call this situation a "liquidity trap."

soared during the period from 2000 to 2007, at their peak they still constituted only about 13.5 percent of the new mortgages issued in the United States.[23] The impact of those defaults was, of course, dramatically amplified by the banks' use of complex financial derivatives. That risk has not been eliminated. A 2014 report by a coalition of bank regulators from the United States and nine other developed countries warned that "five years after the crisis large firms have made only some progress" in addressing the risks associated with derivatives, and that "progress has been uneven and remains, on the whole, unsatisfactory."[24] In other words, the danger that even a localized increase in loan defaults could set off another global crisis remains very real.

The most frightening long-term scenario of all might be if the global economic system eventually manages to adapt to the new reality. In a perverse process of creative destruction, the mass-market industries that currently power our economy would be replaced by new industries producing high-value products and services geared exclusively toward a super-wealthy elite. The vast majority of humanity would effectively be disenfranchised. Economic mobility would become nonexistent. The plutocracy would shut itself away in gated communities or in elite cities, perhaps guarded by autonomous military robots and drones. In other words, we would see a return to something like the feudal system that prevailed during the Middle Ages. There would be one very important difference, however: medieval serfs were essential to the system since they provided the agricultural labor. In a futuristic world governed by automated feudalism, the peasants would be largely superfluous.

The 2013 movie *Elysium*, in which the plutocrats migrate to an Eden-like artificial world in Earth orbit, does a pretty good job of bringing this dystopian vision of the future to life. Even some economists have started to worry about this scenario. Noah Smith, a popular economics blogger, warned in a 2014 post of a possible future in which "a teeming, ragged mass of lumpen humanity teeters on the edge of starvation" outside the gates that protect the elite, and that "unlike the tyrannies of Stalin and Mao, robot-enforced tyranny will be robust to shifts in popular opinion. The rabble may think whatever they please, but the Robot Lords will have the guns. Forever."[25] Even coming from a practitioner of the dismal science, that's pretty bleak.*

Technology and a Graying Workforce

Every industrialized nation has a population that is growing steadily older, and this has led to many predictions of a looming worker shortage as the baby boomers reach retirement age and drop out of the labor force. A 2010 report authored by Barry Bluestone and Mark Melnik of Northeastern University predicts that by 2018, there may be as many as 5 million unfilled jobs in the United States as a direct result of the graying workforce and that "30 to 40 percent of all projected additional jobs in the social sector"—which the authors define as including areas like health care, education, community service, arts, and government—could "go begging unless older workers move into them and make them their encore careers."[26] This is obviously a prediction very much at odds with the argument I've been putting forth in these pages. So which vision of the future is correct? Are we headed toward widespread technological unemployment and even more inequality, or will wages finally begin to rise again as employers scramble to find working-age people to fill available jobs?

The impact of retiring workers in the United States is fairly mild compared to the genuine demographic crises faced by many other

---

* In *Elysium,* the rabble eventually infiltrates the elite orbital fortress by hacking into its systems. That's at least one hopeful note regarding this scenario: the elite would have to be very careful about whom they trusted to design and manage their technology. Hacking and cyber attack would likely be the greatest dangers to their continued rule.

advanced countries, especially Japan. If the United States and other advanced countries are indeed headed toward widespread labor shortages, we might expect the problem to become evident in Japan first.

So far, however, the Japanese economy offers very little in the way of evidence for broad-based labor shortfalls. There are certainly shortages in specific areas, most notably for poorly paid elder-care workers, and the government has also expressed concern about a possible shortage of skilled construction workers as the country begins preparation for the 2020 Olympics in Tokyo. However, if workers were generally in short supply, the result ought to be increased wages across the board, and there is really no evidence for this. Since its real estate and stock market crash in 1990, Japan has experienced two decades of stagnation and even outright deflation. Rather than generating jobs that go begging, the economy has produced an entire lost generation of young people—referred to as "freeters"—who have been unable to find stable career paths and often live with their parents well into their thirties and even forties. In February 2014, the Japanese government announced that 2013 base wages, adjusted for inflation, had actually fallen about 1 percent, matching a sixteen-year low that occurred following the 2008 financial crisis.[27]

Generalized labor shortages are even harder to find elsewhere. As of January 2014, the youth unemployment rates in two of Europe's most rapidly graying countries, Italy and Spain, were both at catastrophic levels: 42 percent in Italy and a stunning 58 percent in Spain.[28] While those extraordinary numbers are, of course, a direct result of the financial crisis, one is nonetheless left to wonder just how long we have to wait before the promised labor shortages begin to put a dent in unemployment among younger workers.

I think that one of the most important lessons we should take from Japan echoes the point I have been making throughout this chapter: *workers are also consumers*. As individuals age, they eventually leave the workforce, but they also tend to consume less, and their spending skews more and more toward health care. So while the number of available workers may decrease, demand for products and services also declines, and that means fewer jobs. In other words, the impact of retiring workers may turn out to largely be a wash, and as seniors reduce their spending in line with their falling incomes, that may well become yet another important reason to question whether economic growth will be sustainable. Indeed, in those countries—such as Japan, Poland, and Russia—where the population is actually in decline, it seems likely that long-term economic stagnation or even contraction will be difficult to avoid since population is a critical determinant of the size of an economy.

Even in the United States, where the population continues to grow, there are good reasons to worry that demographics will depress consumer spending. The transition from traditional pensions to defined contribution (401k) plans has left a great many US households in very fragile circumstances as they approach retirement. In an analysis published in February 2014, MIT economist James Poterba found that a remarkable 50 percent of American households aged sixty-five to sixty-nine have retirement account balances of $5,000 or less.[29] According to Poterba's paper, even a household with $100,000 in retirement savings would receive a guaranteed income of only about $5,400 per year (or $450 per month) with no cost-of-living increases, if the entire balance were used to purchase a fixed annuity.[30] In other words, a great many Americans are likely to end up depending almost entirely on Social Security. In 2013, the average monthly Social Security payment was about $1,300, with some retirees receiving as little as $804. These are not incomes that will support robust consumption, especially given that Medicare premiums currently amounting to about $150 per month (and likely to increase) are deducted.

As in Japan, there are sure to be worker shortages in specific areas, especially those tied directly to the aging trend. Recall from Chapter 6 that the Bureau of Labor Statistics projects about 1.8

million new jobs by 2022 in elder-care-related areas like nursing and personal-care aids. However, if you juxtapose that figure against the 2013 research done by Carl Benedikt Frey and Michael A. Osborne of the University of Oxford, suggesting that jobs comprising about 47 percent of total US employment—roughly 64 million jobs—have the potential to be automated within "perhaps a decade or two,"[31] it seems very difficult to argue that we are headed toward a significant overall shortage of workers. Indeed, rather than counteracting the impact of technology, the aging trend coupled with rising inequality may well be poised to significantly undermine consumer spending. Weak demand could then unleash a secondary wave of job losses affecting even those occupations not directly susceptible to automation.*

Consumer Demand in China and Other Emerging Economies

As inequality and demographics combine to dampen consumer spending in the United States, Europe, and other advanced nations, it seems reasonable to expect that consumers in rapidly growing developing countries will help to pick up the slack. These hopes are directed especially at China, where astonishing growth has led to many predictions that the Chinese economy will become the world's largest, perhaps within the next decade or so.

I think there are a number of reasons to be skeptical about the idea that China and the rest of the emerging world will become primary drivers of global consumer demand anytime soon. The first problem is that China faces a massive demographic shock of its own.

---

* For example, waiting tables in a full-service restaurant would require a very advanced robot—something that we're unlikely to see anytime soon. However, when consumers are struggling, restaurant meals are one of the first things to go, so waiters would still be at risk.

The country's one-child policy has been successful in limiting population growth, but it has also resulted in a rapidly aging society. By 2030, there will be well over 200 million senior citizens in China, roughly double the number in 2010. More than a quarter of the country's population will be sixty-five or older—and more than 90 million people will be at least eighty—by 2050.[32] The rise of capitalism in China resulted in the demise of the "iron rice bowl," under which state-owned industries provided pensions. Retirees now have to fend largely for themselves or rely on their children, but the collapsing fertility rate has led to the infamous "1-2-4" problem in which a single working-age adult will eventually have to help support two parents and four grandparents.

The lack of a social safety net for older citizens is probably one important driver of China's astonishingly high savings rate, which has been estimated to be as much as 40 percent. The high cost of real estate relative to incomes is another important factor. Many workers routinely save more than half their incomes in the hope of someday putting together the down payment for a home.[33]

Households that are stashing away such an enormous share of their incomes are obviously not doing a lot of spending, and indeed, personal consumption amounts to only about 35 percent of China's economy—roughly half the level in the United States. Instead, Chinese economic growth has been powered primarily by manufacturing exports together with an astonishingly high level of investment. In 2013, the share of China's GDP attributable to investment in things like factories, equipment, housing, and other physical infrastructure surged to 54 percent, up from about 48 percent a year earlier.[34] Nearly everyone agrees that this is fundamentally unsustainable. After all, investments have to eventually pay for themselves, and that happens as a result of consumption: factories have to produce goods that are profitably sold, new housing has to be rented, and so forth. The need for China to restructure its economy in favor of domestic spending has been acknowledged by the government and widely discussed for years, and yet virtually no tangible progress has been made. Googling the phrase "China rebalancing" brings up more than 3 million web pages, nearly all of which, I suspect, say roughly the same thing: China's consumers need to get with the program and start buying stuff.

The problem is that making that happen requires dramatically raising household incomes, as well as addressing the issues that have caused the savings rate to soar. Initiatives such as improving the pension and health care systems may help somewhat by reducing the financial risks faced by households. The Chinese central bank has also recently announced plans to relax regulations that hold down the interest rate paid on savings accounts. This might turn out to be a dual-edged sword, on the one hand raising the income going to households but on the other further increasing the incentive to save. Allowing deposit rates to rise could also threaten the solvency of many Chinese banks, which now profit from artificially low interest rates.[35] Some factors behind the Chinese propensity to save may be very hard to address. Economists Shang-Jin Wei and Xiaobo Zhang have proposed that the high saving rate may be attributable to the sex imbalance resulting from China's one-child policy. Because women are scarce, the marriage market is very competitive, and men often have to accumulate substantial wealth or own a home in order to attract a potential spouse.[36] It is also quite possible that a strong desire to save is simply an integral aspect of Chinese culture.

It's often remarked that China faces the danger of growing old before it grows rich, but what I think is less generally acknowledged is that China is in a race not just with demographics but also with technology. As we saw in Chapter 1, Chinese factories are already moving aggressively to introduce robots and automation. Some factories are reshoring to advanced countries or moving to even lower-wage countries like Vietnam. A look back at Figure 2.8 in Chapter 2 shows clearly that advancing technology resulted in a relentless sixty-year collapse in American manufacturing employment. It's inevitable that China must ultimately follow essentially the same path, and it's quite possible that the decline in factory employment may turn out to be even more rapid than in the United States. While automation in American factories progressed only as fast as the new technology could be invented, China's manufacturing sector can, in many cases, simply import leading-edge technology from abroad.

In order to negotiate this transition without a surge in unemployment, China will have to employ an ever-increasing fraction of its workforce in the service sector. However, the typical path followed by advanced nations has been to first become wealthy on the basis of a strong manufacturing sector and then make the transition to a service economy. As incomes rise, households typically spend a larger fraction of their incomes on services, thereby helping to create jobs outside the factory sector. The United States had the luxury of building a strong middle class during its "Goldilocks" period following World War II, when technology was progressing rapidly, but still fell far short of substituting completely for workers. China is faced with performing a similar feat in the robotic age—when machines and software will increasingly threaten jobs not just in manufacturing but also in the service sector itself.

Even if China does succeed in rebalancing its economy toward domestic consumption, it seems optimistic to expect that the country's consumer markets will be fully open to foreign companies. In the United States, the financial and business elite profited enormously from globalization; the most politically influential sector of society had a powerful incentive to keep imports flowing. In China, the situation is quite different. The country's elite are more often than not affiliated directly with the government, and their primary concern is keeping the regime in power. The specter of mass unemployment and social unrest is perhaps their greatest fear. There is little doubt that they would choose to implement overtly protectionist policies if faced with that prospect.

The challenges faced by China are even more daunting for poorer countries, which are much further behind in the race against technology. As even the most labor-intensive areas of manufacturing begin to incorporate more automation, the historical path to prosperity may be poised to largely evaporate for these nations. According to one study, about 22 million factory jobs disappeared worldwide between 1995 and 2002. Over the same seven-year period, manufacturing output increased 30 percent.[37] It is not at all clear how the poorest countries in Asia and Africa will manage to dramatically improve their prospects in a world that no longer needs untold millions of low-wage factory workers.

As ADVANCING TECHNOLOGY CONTINUES to drive inequality in both income and consumption, it is poised to eventually undermine the vibrant and broad-based market demand that is essential for continued prosperity. Consumer markets play a critical role not just in supporting current economic activity but also in advancing the overall process of innovation. While individuals or teams generate new ideas, it is ultimately consumer markets that create the incentive for innovation. Consumers also determine which new ideas succeed—and which are destined to fail. This "wisdom of crowds" function is essential to the Darwinian process through which the best innovations rise above the rest and ultimately scale across the economy and society.

While there's a commonly held belief that business investment is focused on the longer term future and largely independent of current consumption, historical data shows this to be a myth. In virtually every US recession since the 1940s, investment has fallen precipitously.[38] The investment decisions that businesses make are deeply influenced by both the current economic environment and the near-term outlook. In other words, tepid consumer demand today can rob us of prosperity in the future.

In an environment where consumers continue to struggle, many businesses will be inclined to focus on cutting costs rather than expanding markets. One of the few relative bright spots for potential investment is likely to be labor-saving technology. Venture capital and research-and-development investment might then flow disproportionately into innovations specifically geared toward eliminating workers or deskilling jobs. At some point down the line, we could end up with plenty of job-seeking robots—but less of the broad-based innovation that improves the overall quality of our lives.

The trends we've examined in this chapter are all based on what I would characterize as a very realistic, and even conservative, view of the way technology is likely to progress. There can be little doubt that those occupations that primarily involve the execution of tasks that are relatively routine and predictable are going to be highly susceptible to further automation over the course of the next decade or so. As these technologies improve over time, more and more jobs will be impacted.

There is an even more extreme possibility, however. A great many technologists—some of whom are considered to be leaders in their fields—have a far more aggressive view of what will ultimately be possible. In the next chapter, we'll take a balanced look at some of these truly advanced, and far more speculative, technologies. It may well be that these breakthroughs will remain science fiction for the foreseeable future—but if they are ultimately realized, that would dramatically amplify the risk of soaring technological unemployment and income inequality, and perhaps lead to scenarios even more dangerous than the economic risks we've focused on so far.

Chapter 9

SUPER-INTELLIGENCE AND THE SINGULARITY

In May 2014, Cambridge University physicist Stephen Hawking penned an article that set out to sound the alarm about the dangers of rapidly advancing artificial intelligence. Hawking, writing in the UK's *The Independent* along with co-authors who included Max Tegmark and Nobel laureate Frank Wilczek, both physicists at MIT, as well as computer scientist Stuart Russell of the University of California, Berkeley, warned that the creation of a true thinking machine "would be the biggest event in human history." A computer that exceeded human-level intelligence might be capable of "outsmarting financial markets, out-inventing human researchers, out-manipulating human leaders, and developing weapons we cannot even understand." Dismissing all this as science fiction might well turn out to be "potentially our worst mistake in history."[1]

All the technology I've described thus far—robots that move boxes or make hamburgers, algorithms that create music, write reports, or trade on Wall Street—employ what is categorized as specialized or "narrow" artificial intelligence. Even IBM's Watson, perhaps the most impressive demonstration of machine intelligence to date, doesn't come close to anything that might reasonably be compared to general, human-like intelligence. Indeed, outside the realm of science fiction, all functional artificial intelligence technology is, in fact, narrow AI.

One of the primary arguments I've put forth here, however, is that the specialized nature of real-world AI doesn't necessarily represent an impediment to the ultimate automation of a great many jobs. The tasks that occupy the majority of the workforce are, on some level, largely routine and predictable. As we've seen, rapidly improving specialized robots or machine learning algorithms that churn through reams of data will eventually threaten enormous numbers of occupations at a wide range of skill levels. None of this requires machines that can think like people. A computer doesn't need to replicate the entire spectrum of your intellectual capability in order to displace you from your job; it only needs to do the specific things you are paid to do. Indeed, most AI research and development, and nearly all venture capital, continue to be focused on specialized applications, and there's every reason to expect these technologies to become dramatically more powerful and flexible over the coming years and decades.

Even as these specialized undertakings continue to produce practical results and attract investment, a far more daunting challenge lurks in the background. The quest to build a genuinely intelligent system—a machine that can conceive new ideas, demonstrate an awareness of its own existence, and carry on coherent conversations—remains the Holy Grail of artificial intelligence.

Fascination with the idea of building a true thinking machine traces its origin at least as far back as 1950, when Alan Turing published the paper that ushered in the field of artificial intelligence. In the decades that followed, AI research was subjected to a boom-and-bust cycle in which expectations repeatedly soared beyond any realistic technical foundation, especially given the speed of the computers available at the time. When disappointment inevitably followed, investment and research activity collapsed and long, stagnant periods that have come to be called "AI winters" ensued. Spring has once again arrived, however. The extraordinary power of today's computers combined with advances in specific areas of AI research, as well as in our understanding of the human brain, are generating a great deal of optimism.

James Barrat, the author of a recent book on the implications of advanced AI, conducted an informal survey of about two hundred researchers in human-level, rather than merely narrow, artificial intelligence. Within the field, this is referred to as Artificial General Intelligence (AGI). Barrat asked the computer scientists to select from four different predictions for when AGI would be achieved. The results: 42 percent believed a thinking machine would arrive by 2030, 25 percent said by 2050, and 20 percent thought it would happen by 2100. Only 2 percent believed it would never happen. Remarkably, a number of respondents wrote comments on their surveys suggesting that Barrat should have included an even earlier option—perhaps 2020.[2]

Some experts in the field worry that another expectations bubble might be building. In an October 2013 blog post, Yann LeCun, the director of Facebook's newly created AI research lab in New York City, warned that "AI 'died' about four times in five decades because of hype: people made wild claims (often to impress potential investors or funding agencies) and could not deliver. Backlash ensued."[3] Likewise, NYU professor Gary Marcus, an expert in cognitive science and a blogger for the *New Yorker,* has argued that recent breakthroughs in areas like deep learning neural networks, and even some of the capabilities attributed to IBM Watson, have been significantly over-hyped.[4]

Still, it seems clear that the field has now acquired enormous momentum. In particular, the rise of companies like Google, Facebook, and Amazon has propelled a great deal of progress. Never before have such deep-pocketed corporations viewed artificial intelligence as absolutely central to their business models—and never before has AI research been positioned so close to the nexus of competition between such powerful entities. A similar competitive dynamic is unfolding among nations. AI is becoming indispensable to militaries, intelligence agencies, and the surveillance apparatus in authoritarian states.* Indeed, an all-out AI arms race might well be looming in the near future. The real question, I think, is not whether the field as a whole is in any real danger of another AI winter but, rather, whether progress remains limited to narrow AI or ultimately expands to Artificial General Intelligence as well.

If AI researchers do eventually manage to make the leap to AGI, there is little reason to believe that the result will be a machine that simply matches human-level intelligence. Once AGI is achieved, Moore's Law alone would likely soon produce a computer that exceeded human intellectual capability. A thinking machine would, of course, continue to enjoy all the advantages that computers currently have, including the ability to calculate and access information at speeds that would be incomprehensible for us. Inevitably, we would soon share the planet with something entirely unprecedented: a genuinely alien—and superior—intellect.

And that might well be only the beginning. It's generally accepted by AI researchers that such a system would eventually be driven to direct its intelligence inward. It would focus its efforts on improving its own design, rewriting its software, or perhaps using evolutionary programming techniques to create, test, and optimize enhancements to its design. This would lead to an iterative process of "recursive improvement." With each revision, the system would become smarter

* Given recent developments, some readers may be tempted to inject a somewhat snide remark about the National Security Agency at this point. As Hawking's article suggests, there are genuine (and conceivably existential) dangers associated with artificial intelligence. If truly advanced AI is destined to arise somewhere, the NSA is far from the least attractive option.

and more capable. As the cycle accelerated, the ultimate result would be an "intelligence explosion"—quite possibly culminating in a machine thousands or even millions of times smarter than any human being. As Hawking and his collaborators put it, it "would be the biggest event in human history."

If such an intelligence explosion were to occur, it would certainly have dramatic implications for humanity. Indeed, it might well spawn a wave of disruption that would scale across our entire civilization, let alone our economy. In the words of futurist and inventor Ray Kurzweil, it would "rupture the fabric of history" and usher in an event—or perhaps an era—that has come to be called "the Singularity."

The Singularity

The first application of the term "singularity" to a future technology-driven event is usually credited to computer pioneer John von Neumann, who reportedly said sometime in the 1950s that "ever accelerating progress . . . gives the appearance of approaching some essential singularity in the history of the race beyond which human affairs, as we know them, could not continue."[5] The theme was fleshed out in 1993 by San Diego State University mathematician Vernor Vinge, who wrote a paper entitled "The Coming Technological Singularity." Vinge, who is not given to understatement, began his paper by writing that "[w]ithin thirty years, we will have the technological means to create superhuman intelligence. Shortly after, the human era will be ended."[6]

In astrophysics, a singularity refers to the point within a black hole where the normal laws of physics break down. Within the black hole's boundary, or event horizon, gravitational force is so intense that light itself is unable to escape its grasp. Vinge viewed the technological singularity in similar terms: it represents a discontinuity in human progress that would be fundamentally opaque until it occurred. Attempting to predict the future beyond the Singularity would be like an astronomer trying to see inside a black hole.

The baton next passed to Ray Kurzweil, who published his book *The Singularity Is Near: When Humans Transcend Biology* in 2005. Unlike Vinge, Kurzweil, who has become the Singularity's primary evangelist, has no qualms about attempting to peer beyond the event horizon and give us a remarkably detailed account of what the future will look like. The first truly intelligent machine, he tells us, will be built by the late 2020s. The Singularity itself will occur some time around 2045.

Kurzweil is by all accounts a brilliant inventor and engineer. He has founded a series of successful companies to market his inventions in areas like optical character recognition, computer-generated speech, and music synthesis. He's been awarded twenty honorary doctorate degrees as well as the National Medal of Technology and was inducted into the US Patent Office's Hall of Fame. *Inc.* magazine once referred to him as the "rightful heir" to Thomas Edison.

His work on the Singularity, however, is an odd mixture composed of a well-grounded and coherent narrative about technological acceleration, together with ideas that seem so speculative as to border on the absurd—including, for example, a heartfelt desire to resurrect his late father by gathering DNA from the gravesite and then regenerating his body using futuristic nanotechnology. A vibrant community, populated with brilliant and often colorful characters, has coalesced around Kurzweil and his ideas. These "Singularians" have gone so far as to establish their own educational institution. Singularity University, located in Silicon Valley, offers unaccredited graduate-level programs focused on the study of exponential technology and counts Google, Genentech, Cisco, and Autodesk among its corporate sponsors.

Among the most important of Kurzweil's predictions is the idea that we will inevitably merge with the machines of the future. Humans will be augmented with brain implants that dramatically enhance intelligence. Indeed, this intellectual amplification is seen as essential if we are to understand and maintain control of technology beyond the Singularity.

Perhaps the most controversial and dubious aspect of Kurzweil's post-Singularity vision is the emphasis that its adherents place on the looming prospect of immortality. Singularians, for the most part, do not expect to die. They plan to accomplish this by achieving a kind of "longevity escape velocity"—the idea being that if you can consistently stay alive long enough to make it to the next life-prolonging innovation, you can conceivably become immortal. This might be achieved by using advanced technologies to preserve and augment your biological body—or it might happen by uploading your mind into some future computer or robot. Kurzweil naturally wants to make sure that he's still around when the Singularity occurs, and so he takes as many as two hundred different pills and supplements every day and receives others through regular intravenous infusions. While it's quite common for health and diet books to make outsized promises, Kurzweil and his physician co-author Terry Grossman take things to an entirely new level in their books *Fantastic Voyage: Live Long Enough to Live Forever* and *Transcend: Nine Steps to Living Well Forever.*

It's not lost on the Singularity movement's many critics that all this talk of immortality and transformative change has deeply religious overtones. Indeed, the whole idea has been derided as a quasi-religion for the technical elite and a kind of "rapture for the nerds." Recent attention given to the Singularity by the mainstream media, including a 2011 cover story in *Time,* has led some observers to worry about its eventual intersection with traditional religions. Robert Geraci, a professor of religious studies at Manhattan College, wrote in an essay entitled "The Cult of Kurzweil" that if the movement achieves traction with the broader public, it "will present a serious challenge to traditional religious communities, whose own promises of salvation may appear weak in comparison."[7] Kurzweil, for his part, vociferously denies any religious connotation and argues that his predictions are based on a solid, scientific analysis of historical data.

The whole concept might be easy to dismiss completely were it not for the fact that an entire pantheon of Silicon Valley billionaires have demonstrated a very strong interest in the Singularity. Both Larry Page and Sergey Brin of Google and PayPal co-founder (and Facebook investor) Peter Thiel have associated themselves with the subject. Bill Gates has likewise lauded Kurzweil's ability to predict the future of artificial intelligence. In December 2012 Google hired Kurzweil to direct its efforts in advanced artificial intelligence research, and in 2013 Google spun off a new biotechnology venture named Calico. The new company's stated objective is to conduct research focused on curing aging and extending the human lifespan.

My own view is that something like the Singularity is certainly possible, but it is far from inevitable. The concept seems most useful when it is stripped of extraneous baggage (like assumptions about immortality) and instead viewed simply as a future period of dramatic technological acceleration and disruption. It might turn out that the essential catalyst for the Singularity—the invention of super-intelligence—ultimately proves impossible or will be achieved only in the very remote future.* A number of top researchers with expertise in brain science have expressed this view. Noam Chomsky, who has studied cognitive science at MIT for more than sixty years, says we're "eons away" from building human-level machine intelligence, and that the Singularity is "science fiction."[8] Harvard

* It's worth noting that, while machine-based intelligence is the most often cited path to super-intelligence, it could also be biologically based. Human intelligence might be augmented with technology, or future humans might be genetically engineered for superior intelligence. While most Western countries would likely be very squeamish about anything with echoes of eugenics, there is evidence that the Chinese have few qualms about the idea. The Beijing Genomics Institute has collected thousands of DNA samples from people known to have very high IQs and is working on isolating the genes associated with intelligence. The Chinese might be able to use this information to screen embryos for high intelligence and drive their population to become smarter over time.

psychologist Steven Pinker agrees, saying, "There is not the slightest reason to believe in a coming singularity. The fact that you can visualize a future in your imagination is not evidence that it is likely or even possible."[9] Gordon Moore, whose name seems destined to be forever associated with exponentially advancing technology, is likewise skeptical that anything like the Singularity will ever occur.[10]

Kurzweil's timeframe for the arrival of human-level artificial intelligence has plenty of defenders, however. MIT physicist Max Tegmark, one of the co-authors of the Hawking article, told *The Atlantic*'s James Hamblin that "this is very near-term stuff. Anyone who's thinking about what their kids should study in high school or college should care a lot about this."[11] Others view a thinking machine as fundamentally possible, but much further out. Gary Marcus, for example, thinks strong AI will take at least twice as long as Kurzweil predicts, but that "it's likely that machines will be smarter than us before the end of the century—not just at chess or trivia questions but at just about everything, from mathematics and engineering to science and medicine."[12]

In recent years, speculation about human-level AI has shifted increasingly away from a top-down programming approach and, instead, toward an emphasis on reverse engineering and then simulating the human brain. There's a great deal of disagreement about the viability of this approach, and about the level of detailed understanding that would be required before a functional simulation of the brain could be created. In general, computer scientists are more likely to be optimistic, while those with backgrounds in the biological sciences or psychology are often more skeptical. University of Minnesota biologist P. Z. Myers has been especially critical. In a scathing blog post written in response to Kurzweil's prediction that the brain will be successfully reverse engineered by 2020, Myers said that Kurzweil is "a kook" who "knows nothing about how the brain works" and has a penchant for "making up nonsense and making ridiculous claims that have no relationship to reality."[13]

That may be beside the point. AI optimists argue that a simulation does not need to be faithful to the biological brain in every detail. Airplanes, after all, do not flap their wings like birds. Skeptics would likely reply that we are nowhere near understanding the aerodynamics of intelligence well enough to build any wings—flapping or not. The optimists might then retort that the Wright brothers built their airplane by relying on tinkering and experimentation, and certainly not on the basis of aerodynamic theory. And so the argument goes.

The Dark Side

While Singularians typically have a relentlessly optimistic outlook regarding the prospect of a future intelligence explosion, others are far more wary. For many experts who have thought deeply about the implications of advanced AI, the assumption that a completely alien and super-human intelligence would, as a matter of course, be driven to turn its energies toward the betterment of humanity comes across as hopelessly naive. The concern among some members of the scientific community is so high that they have founded a number of small organizations focused specifically on analyzing the dangers associated with advanced machine intelligence or conducting research into how to build "friendliness" into future AI systems.

In his 2013 book *Our Final Invention: Artificial Intelligence and the End of the Human Era*, James Barrat describes what he calls the "busy child scenario."[14] In some secret location—perhaps a government research lab, Wall Street firm, or major corporation in the IT industry—a group of computer scientists looks on as an emergent machine intelligence approaches and then exceeds human-level capability. The scientists have previously provided the AI-child with vast troves of information, including perhaps nearly every book ever written as well as data scoured from the Internet. As the system approaches human-level intelligence, however, the researchers disconnect the rapidly improving AI from the outside world. In effect, they lock it in a box. The question is whether it would stay there. After all, the AI might well desire to escape its cage and expand its horizons. To accomplish this, it might use its superior capability to deceive the scientists or to make promises or threats directed at the group as a whole or at particular individuals. The machine would not only be smarter—it would be able to conceive and evaluate ideas and options at an incomprehensible speed. It would be like playing chess against Garry Kasparov, but with the added burden of unfair rules: whereas you have fifteen seconds to make a move, he has an hour. In the view of those scientists who worry about this type of scenario, the risk that the AI might somehow manage to escape its box, accessing the Internet and perhaps copying all or portions of itself onto other computers, is unacceptably high. If the AI were to break out, it could obviously threaten any number of critical systems, including the financial system, military control networks, and the electrical grid and other energy infrastructure.

The problem, of course, is that all of this sounds remarkably close to the scenarios sketched out in popular science fiction movies and novels. The whole idea is anchored so firmly in fantasy that any attempt at serious discussion becomes an invitation for ridicule. It is not hard to imagine the derision likely to be heaped on any major public official or politician who raised such concerns.

Behind the scenes, however, there can be little doubt that interest in AI of all types within the military, security agencies, and major corporations will only grow. One of the obvious implications of a potential intelligence explosion is that there would be an overwhelming first-mover advantage. In other words, whoever gets there first will be effectively uncatchable. This is one of the primary reasons to fear the prospect of a coming AI arms race. The magnitude of that first-mover advantage also makes it very likely that any emergent AI would quickly be pushed toward self-improvement—if not by the system itself, then by its human creators. In this sense, the intelligence explosion might well be a self-fulfilling prophesy. Given this,

I think it seems wise to apply something like Dick Cheney's famous "1 percent doctrine" to the specter of advanced artificial intelligence: the odds of its occurrence, at least in the foreseeable future, may be very low—but the implications are so dramatic that it should be taken seriously.

Even if we dismiss the existential risks associated with advanced AI and assume that any future thinking machines will be friendly, there would still be a staggering impact on the job market and economy. In a world where affordable machines match, and likely exceed, the capability of even the smartest humans, it becomes very difficult to imagine who exactly would be left with a job. In most areas, no amount of education or training—even from the most elite universities—would make a human being competitive with such machines. Even occupations that we might expect to be reserved exclusively for people would be at risk. For example, actors and musicians would have to compete with digital simulations that would be imbued with genuine intelligence as well as super-human talent. They might be newly created personalities, designed for physical perfection, or they might be based on real people—either living or dead.

In essence, the advent of widely distributed human-level artificial intelligence amounts to the realization of the "alien invasion" thought experiment I described in the previous chapter. Rather than primarily being a threat to relatively routine, repetitive, or predictable tasks, machines would now be able to do nearly everything. That would mean, of course, that virtually no one would be able to derive an income from work. Income from capital—or, in effect, from ownership of the machines—would be concentrated into the hands of a tiny elite. Consumers wouldn't have sufficient income to purchase the output created by all the smart machines. The result would be a dramatic amplification of the trends we've seen throughout these pages.

That wouldn't necessarily represent the end of the story, however. Both those who believe in the promise of the Singularity and those who worry about the dangers associated with advanced artificial intelligence often view AI as intertwining with, or perhaps enabling, another potentially disruptive technological force: the advent of advanced nanotechnology.

Advanced Nanotechnology

Nanotechnology is hard to define. From its inception, the field has been poised somewhere on the border between reality-based science and what many would characterize as pure fantasy. It has been subject to an extraordinary degree of hype, controversy, and even outright dread, and has been the focus of multibillion-dollar political battles, as well as a war of words and ideas between some of the top luminaries in the field.

The fundamental ideas that underlie nanotechnology trace their origin back at least to December 1959, when the legendary Nobel laureate physicist Richard Feynman addressed an audience at the California Institute of Technology. Feynman's lecture was entitled "There's Plenty of Room at the Bottom" and in it he set out to expound on "the problem of manipulating and controlling things on a small scale." And by "small" he meant *really* small. Feynman declared that he was "not afraid to consider the final question as to whether, ultimately—in the great future—we can arrange the atoms the way we want; the very atoms, all the way down!" Feynman clearly envisioned a kind of mechanized approach to chemistry, arguing that nearly any substance could be synthesized simply by putting "the atoms down where the chemist says, and so you make the substance."[15]

In the late 1970s, K. Eric Drexler, then an undergraduate at the Massachusetts Institute of Technology, picked up Feynman's baton and carried it, if not to the finish line, then at least through the next lap. Drexler imagined a world in which nano-scale molecular machines were able to rapidly rearrange atoms, almost instantly transforming cheap and abundant raw material into nearly anything we might want to produce. He coined the term "nanotechnology" and wrote two books on the subject. The first, *Engines of Creation: The Coming Era of Nanotechnology*, published in 1986, achieved popular success and was the primary force that thrust nanotechnology into the public sphere. The book provided a trove of new material for science fiction authors and, by many accounts, inspired an entire generation of young scientists to focus their careers on nanotechnology. Drexler's second book, *Nanosystems: Molecular Machinery, Manufacturing, and Computation*, was a far more technical work based on his doctoral dissertation at MIT, where he was awarded the first PhD ever granted in molecular nanotechnology.

The very idea of molecular machines may seem completely farcical until you take in the fact that such devices exist and, in fact, are integral to the chemistry of life. The most prominent example is the ribosome—essentially a molecular factory contained within cells that reads the information encoded in DNA and then assembles the thousands of different protein molecules that form the structural and functional building blocks of all biological organisms. Still, Drexler was making a radical claim, suggesting that such tiny machines might someday move beyond the realm of biology—where molecular assemblers operate in a soft, water-filled environment—and into the world now occupied by macro-scale machines built from hard, dry materials like steel and plastic.

However radical Drexler's ideas were, by the turn of the millennium nanotechnology had clearly entered the mainstream. In 2000, Congress passed, and President Clinton signed, a bill creating the National Nanotechnology Initiative (NNI), a program designed to coordinate investment in the field. The Bush administration followed up in 2004 with the "21st Century Nanotechnology Research and Development Act," which authorized another $3.7 billion. All told, between 2001 and 2013 the US federal government funneled nearly $18 billion into nanotechnology research, through the NNI. The Obama administration requested an additional $1.7 billion for 2014.[16]

While all this seemed fantastic news for research into molecular manufacturing, the reality turned out to be quite different. According to Drexler's account, at any rate, a massive, behind-the-scenes subterfuge took place even as Congress acted to make funding for nanotechnology research available. In his 2013 book *Radical Abundance: How a Revolution in Nanotechnology Will Change Civilization,* Drexler points out that when the National Nanotechnology Initiative was initially conceived in 2000, the plan explained that "the essence of nanotechnology is the ability to work at the molecular level, atom by atom, to create large structures with fundamentally new molecular organization" and that research would seek to gain "control of structures and devices at atomic, molecular, and supramolecular levels and to learn to efficiently manufacture and use these devices."[17] In other words, the NNI's game plan came straight from Feynman's 1959 lecture, and from Drexler's later work at MIT.

Once the NNI was actually implemented, however, an entirely different vision emerged. In Drexler's words, the newly empowered leaders immediately "purged the NNI's plans of any mention of atoms or molecules in connection with manufacturing and redefined nanotechnology to include anything sufficiently small. Tiny particles were in, atomic precision was out."[18] At least from Drexler's perspective, it was as though the nanotechnology ship had been hijacked by pirates who then proceeded to throw the dynamic molecular machines overboard and sail away with a cargo composed entirely of materials built from tiny, but static, particles. Under the purview of the NNI, virtually all the nanotechnology funding went to research based on relatively traditional techniques in chemistry and materials science; the science of molecular assembly and manufacturing ended up with little or nothing.

A number of factors were behind the sudden shift away from molecular manufacturing. In 2000, Sun Microsystems co-founder Bill Joy wrote an article for *Wired* magazine entitled "Why the Future Doesn't Need Us." In his article, Joy highlighted the possibly existential dangers associated with genetics, nanotechnology, and artificial intelligence. Drexler himself had discussed the possibility of out-of-control, self-replicating molecular assemblers that might use us—and just about everything else—as a kind of feedstock. In *Engines of Creation,* Drexler called it the "gray goo" scenario and noted ominously that it "makes one thing perfectly clear: We cannot afford certain kinds of accidents with replicating assemblers."[19] Joy thought that something of an understatement, writing that "[g]ray goo would surely be a depressing ending to our human adventure on Earth, far worse than mere fire or ice, and one that could stem from a simple laboratory accident."[20] Yet more fuel was thrown on the fire in 2002 when Michael Crichton published his best-selling novel *Prey*—which portrayed swarming clouds of predatory nanobots and opened with an introduction that, once again, quoted passages from Drexler's book.

Public concern over gray goo and feasting nanobots was only part of the problem. Other scientists were beginning to question whether molecular assembly was feasible at all. Most prominent among the skeptics was the late (and aptly named) Richard Smalley, who had won the Nobel Prize in chemistry for his work on nano-scale materials. Smalley had come to the conclusion that molecular assembly and manufacturing, outside the realm of biological systems, was fundamentally at odds with the realities of chemistry. In a public debate with Drexler conducted in the pages of scientific journals, he argued that atoms could not simply be shoved into place using mechanical means; rather, they had to be coaxed into forming bonds, and building molecular machinery capable of achieving this would be impossible. Drexler then accused Smalley of misrepresenting his work, and noted that Smalley himself had once said that "when a scientist says something is possible, they're probably underestimating how long it will take. But if they say it's impossible, they're probably wrong." The debate intensified and became more personal, culminating with Smalley accusing Drexler of having "scared our children" and then concluding that "while our future in the real world will be challenging and there are real risks, there will be no such monster as the self-replicating mechanical nanobot of your dreams."[21]

The nature and magnitude of nanotechnology's future impact will depend in large measure on whether Drexler or Smalley ultimately prove to be correct in their assessment of the feasibility of molecular assembly. If Smalley's pessimism prevails, then nanotechnology will continue to be a field focused primarily on the development of new materials and substances. Dramatic progress in this arena has already occurred, most notably with the discovery and development of carbon nanotubes—structures in which sheets of carbon atoms are rolled into long, hollow threads with an extraordinary range of properties. Carbon nanotube–based materials are potentially a hundred times as strong as steel, while weighing only one-sixth as much.[22] They also offer dramatically enhanced conductivity of both electricity and heat. Carbon nanotubes offer the potential for new lightweight structural materials for cars and aircraft, and may also play an important role in the development of next-generation electronic technologies. Other important advances are occurring in the development of powerful new environmental filtering systems and in medical diagnostic tests and cancer treatments. In 2013, researchers at the Indian Institute of Technology Madras announced a nano-particle-based filtering technology that can provide clean water for a family of five at a cost of just $16 per year.[23] Nano-filters may also eventually provide more effective ways to desalinate ocean water. If nanotechnology follows this path, it will continue to grow in importance, with dramatic benefits flowing to a wide range of applications, including manufacturing, medicine, solar energy, construction, and the environment. The fabrication of nano-materials is, however, a highly capital- and technology-intensive process; accordingly, there are few reasons to expect that the industry will create large numbers of new jobs.

If, on the other hand, Drexler's vision proves to be even partially correct, the eventual impact of nanotechnology may be amplified to a level nearly beyond comprehension. In *Radical Abundance,* Drexler describes what a futuristic fabrication facility equipped to produce large products might look like. In a room about the size of a garage, robotic assembly machines surround a moveable platform. The room's back wall is covered by an array of chambers, each of which is a scaled-down model of the fabrication room. Each chamber, in turn, contains still smaller versions of itself. As the chambers scale down in size, the machinery evolves from normal to micro-size, and then finally to the nano-scale, where individual atoms are arranged into molecules. Once the process is started, fabrication begins at the molecular level and then rapidly scales up as each subsequent level assembles the resulting components. Drexler imagines that a factory like this could produce and assemble a complex product like an automobile within a minute or two. A similar facility would just as easily reverse the process, disassembling finished products into constituent materials that could then be recycled.[24]

Clearly, all this remains in the realm of science fiction for the foreseeable future. Nonetheless, the ultimate realization of molecular assembly would mean the end of the manufacturing industry as we understand it; it would also likely bring about the demise of entire sectors of the economy focused on areas like retail, distribution, and waste management. The global impact on employment would be staggering.

At the same time, of course, manufactured products would become vastly less expensive. In a sense, molecular manufacturing offers the prospect of the digital economy made tangible. It's often said that "information wants to be free." Advanced nanotechnology might allow a similar phenomenon to unfold for material goods. Desktop versions of Drexler's fabricator might someday offer capability similar to the "replicator" used in the television show *Star Trek.* Just as Captain Picard's often-repeated command of "Tea, Earl Grey, Hot" instantly conjures up the proper drink, a molecular fabricator might someday create nearly anything we desire.

Among some techno-optimists, the prospect of molecular manufacturing is associated strongly with the concept of an eventual "post-scarcity" economy in which nearly all material goods are abundant and virtually free. Services are likewise assumed to be provided by advanced AI. In this technological utopia, resource and environmental constraints would be eliminated by universal, molecular recycling and abundant clean energy. The market economy might cease to exist, and (as on *Star Trek*) there would be no need for money. While that may sound like a very inviting scenario, there are a great many details that would need to be fleshed out. Land, for example, would still remain scarce, making it unclear how living space would be allocated in a world largely without jobs, money, or opportunities for most people to advance their station economically. Likewise, it's unclear how the incentives necessary for further progress would be maintained in the absence of a market economy.

Physicist (and *Star Trek* fan) Michio Kaku has said that he thinks a nanotechnology-driven utopia might be a possibility within a hundred years or so.* In the meantime, there are a number of more practical and immediate questions associated with molecular manufacturing. The "grey goo" scenario and other fears regarding self-replication remain very real concerns, as does the potential for deliberately destructive use of the technology. Indeed, molecular assembly, if it were weaponized by an authoritarian regime, might bring about a world order very different from utopia. Drexler warns that while the United States has almost completely turned away from any organized research into molecular manufacturing, the same is not necessarily true of other countries. The United States, Europe, and China all make roughly the same level of investment in nanotechnology research, but the focus of this research might be entirely

* You can watch Michio Kaku discuss the post-scarcity economy in the video "Can Nanotechnology Create Utopia?," available on YouTube.

different within each jurisdiction.[25] As with artificial intelligence, there is the potential for an all-out arms race, and prematurely adopting a defeatist approach toward molecular assembly might be tantamount to unilateral disarmament.

THIS CHAPTER HAS BEEN a fairly radical departure from the more practical and immediate arguments I've been making elsewhere throughout the book. The prospects of true thinking machines, advanced nanotechnology—and, especially, the Singularity—are, to say the least, highly speculative. It may be that none of these things are possible, or they may lie centuries in the future. If any of these breakthroughs are ultimately achieved, however, there can be little doubt that they would dramatically accelerate the trend toward automation and massively disrupt the economy in unforeseen ways.

There is also, to some extent, a kind of paradox associated with the realization of these futuristic technologies. The development of both advanced AI and molecular manufacturing will require enormous investment in research and development. However, long before such genuinely advanced technologies become practical, more specialized forms of AI and robotics are likely to threaten vast numbers of jobs at a variety of skill levels. As we saw in the previous chapter, that could well undermine market demand—and therefore the incentive for further investment in innovation. In other words, the research necessary to achieve Singularity-level technologies might never be funded, and progress could, in effect, become self-limiting.

None of the technologies we looked at in this chapter are necessary to the primary arguments I have been putting forth here; rather, they might be viewed as possible—and dramatic—amplifiers of a relentless technology-driven trend toward greater inequality and rising unemployment. In the next chapter we'll look at some possible policy measures that might help counteract that trend.

Chapter 10

TOWARD A NEW ECONOMIC PARADIGM

In an interview with *CBS News,* the president of the United States was asked if the nation's dire unemployment problem was likely to improve soon. "There's no magic solution," he replied. "To even stand still we have to move very fast." By this, he meant that the economy needs to create tens of thousands of new jobs every month just to keep pace with population growth and prevent the unemployment rate from rising even further. He pointed out that "we have a combination of older workers who have been thrown out of work because of technology and younger people coming in" with too little education. The president proposed a tax cut to stimulate the economy, but he kept returning to the subject of education—in particular, advocating support for programs focused on "vocational education" and "job retraining." The problem, he said, wasn't going to solve itself: "[T]oo many people are coming into the labor market and too many machines are throwing people out."[1]

The president's words capture the conventional—and nearly universal—assumption about the nature of the unemployment problem: more education or more vocational training is always the solution. With the proper training, workers will continuously climb the skills ladder, somehow staying just ahead of the machines. They will do more creative work, more "blue-sky" thinking. There is apparently no limit to what average people can be educated and trained to do—and likewise no limit to the number of high-level jobs the economy can create to absorb all these newly trained workers. Education and retraining, it seems, are a solution that is immutable across time.

For those who hold this view, it is perhaps of little import that the president quoted above was named Kennedy and the date was September 2, 1963. As President Kennedy noted, the unemployment rate at the time was about 5.5 percent, and machines were confined almost exclusively to "taking the place of manual labor." Seven months after the interview took place, the Triple Revolution report would land on a new president's desk. It would be another four years before Dr. King would make his own reference to technology and automation in Washington National Cathedral. In the nearly half-century since then, belief in the promise of education as the universal solution to unemployment and poverty has evolved hardly at all. The machines, however, have changed a great deal.

Diminishing Returns to Education

If we were to draw a graph of the gains from ever-increasing investment in education, it seems very likely that we would end up with something that looks like the S-curves we discussed in Chapter 3. The low-hanging fruit of further education is long behind us. High school graduation rates have leveled off at roughly 75 to 80 percent. Most standardized test scores have shown little or no improvement in recent decades. We are on the flat part of the curve, where continued progress will be at best incremental.

An abundance of evidence suggests that many of the students now attending American colleges are academically unprepared for or, in some cases, simply ill-suited to college-level work. Of these, a large share will fail to graduate but very often will nonetheless walk away with daunting student loan burdens. Of those who do graduate, as many as half will fail to land a job that actually requires a college degree, whatever the job description might say. Overall, about 20 percent of US college graduates are considered overeducated for their current occupation, and average incomes for new college graduates have been in decline for more than a decade. In Europe, where many countries provide students with college educations that are free or nearly so, roughly 30 percent of graduates are overqualified for their jobs.[2] In Canada, the number is about 27 percent.[3] In China, a remarkable 43 percent of the workforce is overeducated.[4]

In the United States, the conventional wisdom tends to put most of the blame on students and educators. College students are said to spend too much time socializing and too little time studying. They are choosing fields with easy classes, rather than graduating with degrees in more rigorous technical fields. Yet, as many as a third of American students who do obtain a degree in engineering, science, or other technical fields fail to find a position that utilizes their educational background.[5]

Steven Brint, a sociologist at the University of California, Riverside, who has written extensively on higher education, argues that US colleges actually graduate students who are relatively well-matched to the available job opportunities. Brint notes that "a few jobs require specialized skills that can only be acquired in technical programs, but most jobs are relatively routine." "Following the directives of supervisors is essential" and "reliability and steady effort are highly valued." He concludes that "dedicated work is not required in college because it will not be required at work. In most jobs, showing up and doing the work is more important than achieving outstanding levels of performance."[6] If you were to purposely set out to describe the characteristics of a job vulnerable to automation, it would be hard to do much better than that.

The reality is that awarding more college degrees does not increase the fraction of the workforce engaged in the professional, technical, and managerial jobs that most graduates would like to land. Instead, the result very often is credential inflation; many occupations that once required only a high school diploma are now open only to those with a four-year college degree, the master's becomes the new bachelor's, and degrees from nonelite schools are devalued. We are running up against a fundamental limit both in terms of the capabilities of the people being herded into colleges and the number of high-skill jobs that will be available for them if they manage to graduate. The problem is that the skills ladder is not really a ladder at all: it is a pyramid, and there is only so much room at the top.

Historically, the job market has always looked like a pyramid in terms of worker skills and capabilities. At the top, a relatively small number of highly skilled professionals and entrepreneurs have been responsible for most creativity and innovation. The vast majority of the workforce has always been engaged in work that is, on some level, relatively routine and repetitive. As various sectors of the economy have mechanized or automated, workers have transitioned from routine jobs in one sector to routine jobs in another. The person who would have worked on a farm in 1900, or in a factory in 1950, is today scanning bar codes or stocking shelves at Walmart. In many cases, this transition has required additional training and upgraded skills, but the work has nonetheless remained essentially routine in nature. So, historically, there has been a reasonable match between the types of work required by the economy and the capabilities of the available workforce.

It's becoming increasingly clear, however, that robots, machine learning algorithms, and other forms of automation are gradually going to consume much of the base of the job skills pyramid. And because artificial intelligence applications are poised to increasingly encroach on more skilled occupations, even the safe area at the top of the pyramid is likely to contract over time. The conventional wisdom is that, by investing in still more education and training, we are going to somehow cram everyone into that shrinking region at the very top.* I think that assuming this is possible is analogous to believing that, in the wake of the mechanization of agriculture, the majority of displaced farm workers would be able to find jobs driving tractors. The numbers simply don't work.

American primary and secondary education, of course, also has major problems. Inner-city high schools have staggering dropout rates, and children in the most poverty-stricken areas are at a significant disadvantage even before they enter the school system. Even if we could wave a magic wand and give every American child a top-notch education, that would only mean more high school graduates entering college and competing for the limited number of jobs at the top of the pyramid. That's not to say we shouldn't wave the wand, of course: we should—but we shouldn't expect it to solve all our problems. Needless to say, the magic wand doesn't exist, and although there is a universal consensus that we need to improve our schools, it exists at only the most superficial level. Start talking about more money for schools, charter schools, firing bad teachers, paying good teachers more, longer school days (or years), or vouchers for private schools, and the situation will rapidly degrade into political intractability.

The Anti-Automation View

Another often-proffered solution is simply to try to put a stop to this relentless progression toward ever more automation. At its most blunt, this might take the form of a union resisting the installation of new machinery in a factory, warehouse, or supermarket. There is also a more nuanced intellectual argument which says that too much automation is simply bad for us—and quite possibly dangerous.

* Keep in mind that many of those higher-skill jobs may also be threatened by offshoring.

Nicholas Carr is perhaps the best-known proponent of this view. In his 2010 book *The Shallows,* Carr argues that the Internet may be having a negative impact on our ability to think. In a 2013 article for *The Atlantic,* entitled "All Can Be Lost: The Risk of Putting Our Knowledge in the Hands of Machines," he makes a similar argument about the impact of automation. Carr complains about the "the rise of 'technology-centered automation' as the dominant design philosophy of computer engineers and programmers" and believes that this "philosophy gives precedence to the capabilities of technology over the interests of people."[7]

Carr's *Atlantic* article includes a number of anecdotes demonstrating how automation can erode human skills, in some cases with disastrous consequences. Some are a bit arcane: for example, Inuit hunters in Northern Canada are losing a 4,000-year-old ability to navigate in a frigid environment as they search for game because they are now relying on GPS. Carr's best examples, however, are drawn from aviation. The paradox of increased cockpit automation is that, while the technology reduces the cognitive burden on pilots and almost certainly contributes to a better overall safety record, it also means that pilots spend less time actively flying the plane. In other words, they get less practice and, over time, the nearly instinctual reactions that professional pilots develop over countless hours of training can begin to degrade. Carr worries that a similar effect is likely to cascade across offices, factories, and other workplaces as automation continues its advance.

This idea that engineering "design philosophy" is the problem has also been embraced to some degree by economists. MIT's Erik Brynjolfsson, for example, has called for a "New Grand Challenge for Entrepreneurs, Engineers and Economists" to "invent complements, not substitute[s] for labor" and "replace [the] labor saving and automation mindset with [a] maker and creator mindset."[8]

Suppose that a start-up company were to rise to Brynjolfsson's challenge and build a system specifically designed to keep people in the loop. A competitor designs a system that is fully automated, or at least requires minimal human intervention. In order for the more people-oriented system to be economically competitive, one of two things has to be true. Either it has be significantly less expensive, to offset the increased labor costs, or it has to produce results so superior that they deliver substantially greater value to customers and ultimately generate enough additional revenue to make those extra costs appear to be a rational investment. There are good reasons to be skeptical that either case would be true in the vast majority of circumstances. In the case of white-collar automation, both systems would be composed primarily of software, so there would be little reason for a major cost differential. It's possible that, in a few areas central to a business's primary focus, the people-oriented system might have a meaningful advantage (and ability to generate more revenue over the long run), but for the majority of more routine operational activities, where simply showing up is more important than doing an outstanding job, this again seems unlikely.

Furthermore, this simple cost comparison very likely understates the bias toward automation. Every new worker a business hires adds to a whole slew of peripheral costs. The more workers you have, the more managers and human resources staff you need. Workers likewise need offices, equipment, and parking spaces. Workers also introduce uncertainty: they get sick, perform poorly, take vacation, have car trouble, quit entirely, and generally run into a myriad of other potential issues.

Every new worker you hire also comes with a serving of potential liability. An employee might get hurt at work—or might somehow harm someone else. There's also the risk of reputational harm to the business. If you want to see some major corporate brands take a beating, try Googling the phrase "delivery driver throws package."

The bottom line is that, despite all the rhetoric about "job creators," rational business owners do *not want* to hire more workers: they hire people only because they have to. The progression toward ever more automation is not an artifact of "design philosophy" or the personal preferences of engineers: it is fundamentally driven by capitalism. The "rise of 'technology-centered automation'" that Carr worries about took place at least two hundred years ago, and the Luddites were unhappy about it. The only difference today is that exponential progress is now pushing us toward the endgame. For any rational business, the adoption of labor-saving technology will almost invariably prove to be irresistible. Changing that would require far more than an appeal to engineers and designers: it would require modifying the basic incentives built into the market economy.

Some of the concerns raised by Carr are real, but the good news is that in the most important areas we already have safeguards in place. The most dramatic examples of automation-related risks are those that threaten lives or lead to potential catastrophe. Aviation comes up again and again. Yet these areas are already subject to extensive regulation. The aviation industry has been aware of the interaction between cockpit automation and pilot skill levels for years and has presumably incorporated this knowledge into its training procedures. There is no question that the overall safety record of the modern aviation system is astonishing. Some technologists foresee aircraft automation taken to the extreme. Sebastian Thrun, for example, recently told the *New York Times* that "airline pilot" would be a "profession of the past" in the not-too-distant-future.[9] I really don't think we will see three hundred people filing onto an airplane with no pilots onboard anytime soon. The combination of regulation, potential liability, and simple acceptance on the part of society is certain to create powerful headwinds in occupations that are directly tied to public safety. It will be the tens of millions of *other jobs*—the fast food workers, the office drones, and all the rest—where the impact of automation on employment is likely to be most dramatic. In these areas, a potential technical failure or an erosion of skill has far less spectacular consequences, and there are relatively few barriers to a relentless progression toward full automation—driven, of course, by market incentives.

Throughout our economy and society, machines are gradually undergoing a fundamental transition: they are evolving beyond their historical role as tools and, in many cases, becoming autonomous workers. Carr views this as dangerous and would presumably like to somehow put a stop to it. The reality, however, is that the astonishing wealth and comfort we have achieved in modern civilization are a direct result of the forward march of technology—and the relentless drive toward ever more efficient ways to economize on human labor has arguably been the single most important factor powering that progress. It's easy to claim that you are against the idea of too much automation, while still not being anti-technology in the general sense. In practice, however, the two trends are inextricably tied together, and anything short of a massive—and certainly ill-advised—intrusion of government into the private sector seems destined to fail at any attempt to halt the inevitable, market-driven rise of autonomous technology in the workplace.

The Case for a Basic Income Guarantee

If we accept the idea that ever more investment in education and training is unlikely to solve our problems, while calls to somehow halt the rise of job automation are unrealistic, then we are ultimately forced to look beyond conventional policy prescriptions. In my view, the most effective solution is likely to be some form of basic income guarantee.

A basic, or guaranteed minimum, income is far from a new idea. In the context of the contemporary American political landscape, a guaranteed income is likely to be disparaged as "socialism" and a massive expansion of the welfare state. The idea's historical origins, however, suggest something quite different. While a basic income has been embraced by economists and intellectuals on both sides of the political spectrum, the idea has been advocated especially forcefully by conservatives and libertarians. Friedrich Hayek, who has become an iconic figure among today's conservatives, was a strong proponent of the idea. In his three-volume work *Law, Legislation and Liberty*, published between 1973 and 1979, Hayek suggested that a guaranteed income would be a legitimate government policy designed to provide insurance against adversity, and that the need for this type of safety net is the direct result of the transition to a more open and mobile society where many individuals can no longer rely on traditional support systems:

> There is, however, yet another class of common risks with regard to which the need for government action has until recently not been generally admitted. . . . The problem here is chiefly the fate of those who for various reasons cannot make their living in the market . . . that is[,] all people suffering from adverse conditions which may affect anyone and against which most individuals cannot alone make adequate protection but in which a society that has reached a certain level of wealth can afford to provide for all.
>
> The assurance of a certain minimum income for everyone, or a sort of floor below which nobody need fall even when he is unable to provide for himself, appears not only to be a wholly legitimate protection against a risk common to all, but a necessary part of the Great Society in which the individual no longer has specific claims on the members of the particular small group into which he was born.[10]

Those words might well come as something of a surprise to those conservatives who buy into the currently fashionable extreme right-wing caricature of Hayek. To be sure, when Hayek uses the words "Great Society" he means something quite different from what Lyndon Johnson envisioned when he used the same phrase. Rather than an ever-expanding welfare state, Hayek saw a society based on individual freedom, market principles, the rule of law, and limited government. Still, his reference to "the Great Society" as well as his recognition that "a society that has reached a certain level of wealth can afford to provide for all" seems to stand in stark contrast to today's more extreme conservative views, which are more likely to embrace Margaret Thatcher's statement that "there is no such thing as society."

Indeed, a proposal for a guaranteed income would today almost certainly be attacked as a liberal mechanism for attempting to bring about "equal outcomes." Hayek himself explicitly rejected this, however, writing that "it is unfortunate that the endeavor to secure a uniform minimum for all who cannot provide for themselves has become connected with the wholly different aims of securing a 'just' distribution of incomes."[11] For Hayek, a guaranteed income had nothing to do with equality or "just distribution"—it was about insurance against adversity as well as efficient social and economic function.

I think one of the primary takeaways from Hayek's view is that he was fundamentally a realist rather than an ideologue. He understood that the nature of society was changing; people had moved from farms, where they were largely self-sufficient, to cities, where they depended on jobs, and extended family structures were breaking down—leaving individuals to assume higher risks. He had no problem with a role for government in helping to insure against those risks. This idea that the role of government can evolve over time is, of course, highly applicable to the challenges we face today.*

The conservative argument for a basic income centers on the fact that it provides a safety net coupled with individual freedom of choice. Rather than having government intrude into personal economic decisions, or get into the business of directly providing products and services, the idea is to give everyone the means to go out and participate in the market. It is fundamentally a market-oriented

* The idea that both government and society must evolve with the times is echoed by another conservative icon. Here's a quote from Thomas Jefferson, which is engraved into panel #4 of the Jefferson Memorial: "I am not an advocate for frequent changes in laws and constitutions, but laws and institutions must go hand in hand with the progress of the human mind. As that becomes more developed, more enlightened, as new discoveries are made, new truths discovered and manners and opinions change, with the change of circumstances, institutions must advance also to keep pace with the times. We might as well require a man to wear still the coat which fitted him when a boy as civilized society to remain ever under the regimen of their barbarous ancestors."

approach to providing a minimal safety net, and its implementation would make other less efficient mechanisms—the minimum wage, food stamps, welfare, and housing assistance—unnecessary.

If we adopt Hayek's pragmatism and apply it to the situation likely to develop in the coming years and decades, it seems very likely that government will ultimately be called upon to take some type of action in the face of the increased risks to individual economic security brought about by advancing technology. If we reject Hayek's market-oriented solution, then we'll inevitably end up with an expansion of the traditional welfare state, along with all the problems that accompany it. It's easy to imagine the eventual rise of vast new bureaucracies geared toward feeding and housing masses of economically disenfranchised people—perhaps in dystopian quasi-institutional environments.

Indeed, this is very likely the path of least resistance—and the default if we simply do nothing. A basic income would be efficient and would have relatively low administrative costs. A bureaucratic expansion of the welfare state would be far more expensive on a per capita basis, and far more unequal in its impact. It would almost certainly help fewer people, but it would create a number of traditional jobs, some of which would be very lucrative. There would also be abundant opportunities for private-sector contractors to jump on the gravy train. These elite beneficiaries—the high-level administrators, the private-company executives—are sure to exert substantial political pressure for things to evolve along this path.

There are, of course, plenty of examples of this kind of thing already. Massive weapons programs that the Pentagon does not want are protected by Congress because they create a small number of jobs (relative to their enormous costs) and pad the profits of large corporations. The United States has a staggering 2.4 million people locked up in jails and prisons—a per capita incarceration rate more than three times that of any other country and more than ten times that of advanced nations like Denmark, Finland, and Japan.

As of 2008, about 60 percent of these people were nonviolent offenders, and the annual per capita cost of housing them was about $26,000.[12] Powerful elites—including, for example, prison guards' unions and executives at the private corporations that operate many prisons—have strong incentives to ensure that the United States remains an extreme outlier in this area.

For progressives, a guaranteed income may be an easier sell in the current political environment. Despite Hayek's argument to the contrary, many liberals would likely embrace the idea as a method to achieve more social and economic justice. A basic income could effectively become a brute-force algorithm designed to alleviate poverty and mitigate income inequality. At a stroke of the presidential pen, extreme poverty and homelessness in the United States might effectively be eradicated.

Incentives Matter

The most important factor in designing a workable guaranteed income scheme is getting the incentives right. The objective should be to provide a universal safety net as well as a supplement to low incomes—but without creating a disincentive to work and to be as productive as possible. The income provided should be relatively minimal: enough to get by, but not enough to be especially comfortable. There is also a strong argument for initially setting the income level even lower than this and then gradually increasing it over time after studying the impact of the program on the workforce.

There are two general approaches to implementing a guaranteed income. An unconditional basic income is paid to every adult citizen regardless of other income sources. Guaranteed minimum incomes (and other variations, such as a negative income tax) are paid only to people at the bottom of the income distribution and are phased out as other income sources rise. While the second alternative is obviously less expensive, it carries with it the danger of disastrous perverse incentives. If the guaranteed income is means-tested at relatively low income levels, recipients will see an effective tax rate on any further earnings that can reach confiscatory levels. In other words, they can fall into a "poverty trap" where there is little or no benefit to working harder. Perhaps the worst possible example of this occurs with the Social Security disability program, which many people likely attempt to utilize as a kind of guaranteed income when their other options are exhausted. Once a person is approved for disability payments, any attempt to work beyond that point carries the danger of losing both the income and the accompanying health care benefits. As a result, virtually no one who gets into the program ever works again.

Clearly, if a guaranteed income is means-tested, then this should happen at a relatively high level, preferably well into middle-class territory. A person who decides to forego other earning opportunities then faces a long fall. Another good idea would be to discriminate between active and passive income. A guaranteed income might be means-tested aggressively against passive income such as a pension, investment income, or Social Security. Active income like wages from a job, self-employment income, or earnings from a small business either would not be means-tested at all or would occur at a much higher level. This should ensure a consistent incentive for everyone to work as hard as possible, given the opportunities available.

A guaranteed income scheme would also be likely to create a number of more subtle incentives for both individuals and families. Conservative social scientist Charles Murray's 2006 book *In Our Hands: A Plan to Replace the Welfare State* argues that a guaranteed income would be likely to make non-college-educated men more attractive marriage partners. This group has been the hardest hit by the impact of both technology and factory offshoring on the job market. A guaranteed income might help increase marriage rates among lower-income groups, while helping to reverse the trend toward more children being raised in single-parent households. It would also, of course, make it more feasible for one parent to choose to stay at home with young children. These are all things that might appeal to people across the political spectrum.

Beyond this, I think there are compelling reasons to go further and build some explicit incentives into a basic income program. The most important of these would be geared toward education, especially at the high school level. Recent data shows that there continues to be a strong economic incentive to pursue a college degree. However, the unfortunate reality is that this is the case not so much because opportunities for college graduates are expanding dramatically but because prospects for those with only a high school diploma are collapsing. I think this creates a real danger that, for a significant number of people who are not destined to graduate from college, the incentive to complete high school may be diminished. If a struggling high school student knows that he will receive a guaranteed income regardless of whether or not he graduates, that obviously creates a very powerful perverse incentive. Therefore, we ought to pay a somewhat higher income to those who earn their high school diploma (or the equivalent through testing).

The general idea is that we should value education as a public good. We all benefit when the people around us are more educated; this generally results in a more civil society as well as a more productive economy. If we are destined to transition into an era where traditional work becomes less available, then an educated population will be in a better position to find constructive uses for leisure time. Technology is creating many opportunities to spend time in productive ways. Wikipedia has been built through countless hours of labor by unpaid contributors. The open source software movement offers another example. Many people start small online businesses to supplement their income. Yet, in order to successfully participate in such activities you need to reach a minimal educational threshold.

Other incentives might also be implemented. For example, a higher income might be paid to those who volunteer for community service activities or participate in environmental projects. When I suggested building explicit incentives of this type into a guaranteed income in my previous book, *The Lights in the Tunnel,* I received a fair amount of pushback from more libertarian readers who strongly objected to the idea of an intrusive "nanny state." Nonetheless, I think there are some basic incentives—most critically education—that nearly everyone ought to be able to agree on. The essential idea is to replicate (albeit artificially) some of the incentives associated with traditional jobs. In an age when more education may not always lead to an improved career path, it's important to ensure that everyone has a compelling reason to at least complete high school. To me, the resulting advantages to society seem obvious. Even Ayn Rand, if she were rational, would presumably perceive a personal benefit in being surrounded by people with a higher level of education and more options for constructive use of their free time.

The Market as a Renewable Resource

Aside from the need to provide a basic safety net, I think there is a powerful economic argument for a guaranteed income. As we saw in Chapter 8, increasing technology-driven inequality is likely to threaten broad-based consumption. As the job market continues to erode and wages stagnate or fall, the mechanism that gets purchasing power into the hands of consumers begins to break down, and demand for products and services suffers.

To visualize the problem, I find it useful to think of markets as renewable resources. Imagine a consumer market as a lake full of fish. When a business sells products or services into the market, it catches fish. When it pays wages to its employees, it tosses fish back into the lake. As automation progresses and jobs disappear, fewer fish get returned to the lake. Again, keep in mind that nearly all major industries are dependent on catching large numbers of moderately sized fish. Increasing inequality will result in a small number of very large fish, but from the point of view of most mass-market industries these aren't worth a whole lot more than normal-sized fish. (The billionaire is not going to buy a thousand smart phones, cars, or restaurant meals.)

This is what's known as a classic "tragedy of the commons" problem. The vast majority of economists would likely agree that a situation like this calls for some kind of government intervention. In the absence of this, there is no individual incentive to do anything except catch as many fish as possible. Real-world fishermen may understand fully that their lake or ocean is being over-fished and that their livelihoods will soon be threatened, but they will nonetheless go out every day and maximize their catch because they know their competitors will do the same. The only viable solution is to have some regulatory authority step in and impose limits.

In the case of our consumer market, we don't want to limit the number of virtual fish that businesses can catch. Instead, we want to make sure the fish get replenished. A guaranteed income is one very effective way to do this. The income gets purchasing power directly into the hands of lower- and middle-income consumers.

If we look further into the future and assume that machines will eventually replace human labor to a substantial degree, then I think some form of direct redistribution of purchasing power becomes essential if economic growth is to continue. In a May 2014 paper on the future of American economic growth, economists John G. Fernald and Charles I. Jones speculated that robots could "increasingly replace labor in the production function for goods." They then go on to suggest that "in the limit, if capital can replace labor entirely, growth rates could explode, with incomes becoming infinite in finite time."[13] This strikes me as a nonsensical result; it's the kind of thing you get by plugging numbers into an equation without really thinking through the implications. If machines substitute for workers entirely, then no one has a job or an income from any type of labor. The vast majority of consumers have no purchasing power. So how can the economy keep growing? Perhaps the tiny percentage of people with significant capital ownership could do all the consuming, but they would need to continuously purchase goods and services of staggering value in order to keep the global economy growing.* And this, of course, is the "techno-feudalism" scenario we looked at in Chapter 8—not an especially hopeful outcome.

There is, however, a more optimistic view. Perhaps the mathematical model Fernald and Jones are using might be said to *assume* a mechanism—other than income from labor—for distributing purchasing power. If something like a guaranteed income were implemented, and if the income were increased over time to support continued economic growth, then the idea that growth could explode and incomes could soar might make sense. This will not happen automatically; the market is not going to sort things out on its own. A fundamental restructuring of our economic rules will be required.

I think that viewing markets—or the entire economy—as a resource also works well from another perspective. Recall that in Chapter 3, I argued that the technologies poised to transform the job market result from a cumulative effort that has spanned generations and has involved countless individuals, and has often been funded by taxpayers. To some extent, you can make a reasonable argument that all these accumulated advances—as well as the economic and political institutions that enable a vibrant market economy—are really a resource that belongs to all citizens. A term often used in place of "guaranteed income" is "citizen's dividend," which I think effectively

---

* What we call "the economy" is really the total value of all the goods and services produced and sold to someone. The economy can either produce enormous numbers of low and moderately priced goods and services, or a much smaller number of very high-value goods and services. The first scenario requires broad distribution of purchasing power; this is currently made possible by jobs. In the second scenario it is unclear what products and services the economy could produce that would be valued so highly by the wealthy elite. Whatever these high-priced goods were, they would need to be consumed voraciously by the lucky few—otherwise the economy would not grow at all: it would contract.

captures the argument that everyone should have at least a minimal claim on a nation's overall economic prosperity.

The Peltzman Effect and Economic Risk Taking

In 1975, the University of Chicago economist Sam Peltzman published a study showing that regulations designed to improve automobile safety had failed to result in a significant reduction in highway fatalities. The reason, he argued, was that drivers simply compensated for the perceived increase in safety by taking more risks.[14]

This "Peltzman effect" has since been demonstrated in a wide range of areas. Children's playgrounds, for example, have become much safer. Steep slides and high climbing structures have been removed and cushioned surfaces have been installed. Yet, studies have shown that there has been no meaningful reduction in playground-related emergency room visits or broken bones.[15] Other observers have noted the same phenomenon with respect to skydiving: the equipment has gotten dramatically better and safer, but the fatality rate remains roughly the same as skydivers compensate with riskier behavior.

The Peltzman effect is typically invoked by conservative economists in support of an argument against increased government regulation. However, I think there is every reason to believe that this risk compensation behavior extends into the economic arena. People who have a safety net will be willing to take on more economic risk. If you have a good idea for a new business, it seems very likely that you would be more willing to quit a secure job and make the leap into entrepreneurship if you knew you had access to a guaranteed income. Likewise, you might decide to leave a safe job that offered you few opportunities for personal growth in order to take a more rewarding but less secure position at a small start-up company. A guaranteed income would offer an economic cushion for all types of entrepreneurial activity, from the person starting an online business, to the "mom and pop" retailer or restaurateur, to the small farmer or rancher facing a drought. In many cases, it might be enough to get small businesses through difficult periods that would otherwise bring about their failure. The bottom line is that, rather than resulting in a nation of slackers, a well-designed guaranteed income has the potential to make the economy more dynamic and entrepreneurial.

Challenges, Downsides, and Uncertainties

A guaranteed income is not without downsides and risks. The most important near-term concern is whether or not a strong disincentive to work would be created. While machines clearly seem destined to take on more and more work over time, there is no question that the economy will remain heavily dependent on human labor for the foreseeable future.

There are, to date, no examples of such a policy having been implemented on a national level. The state of Alaska has paid a modest annual dividend funded by oil revenue since 1976; in recent years, the payments have typically been between $1,000 and $2,000 per person. Both adults and children are eligible, so the amount can be significant for families. In October 2013, proponents of a guaranteed income in Switzerland gathered enough signatures to put a proposal for a remarkably generous unconditional monthly stipend of 2,500 Swiss francs (or about $2,800) on the national ballot, although no date has yet been set for the vote. Small-scale experiments in the United States and Canada have shown a reduction of roughly 5 percent in the number of hours that recipients chose to work; however, these were temporary programs and therefore less likely to influence behavior than a permanent program.[16]

One of the greatest political and psychological barriers to the implementation of a guaranteed income would be simple acceptance of the fact that some fraction of recipients will inevitably take the money and drop out of the workforce. Some people will choose to play video games all day—or, worse, spend the money on alcohol or drugs. Some recipients might pool their incomes, crowding into housing or perhaps even forming "slacker communes." As long as the income is fairly minimal and the incentives are designed correctly, the percentage of people making such choices would likely be very low. In absolute numbers, however, they could be quite significant—and quite visible. All of this, of course, would be very hard to reconcile with the general narrative of the Protestant work ethic. Those opposed to the idea of a guaranteed income would likely have little trouble finding disturbing anecdotes that would serve to undermine public support for the policy.

In general, I think the fact that some people would elect to work less—or perhaps even not at all—should not be viewed in universally negative terms. It's important to keep in mind that the individuals who choose to drop out will be self-selecting. In other words, they will generally be among the least ambitious and industrious members of the population.* In a world where everyone is forced to compete for a dwindling number of jobs, there is no reason to believe that the most productive people will always be the ones to land those jobs. If some people work less or drop out entirely, then wages for those who are willing to work hard may rise somewhat. That fact that incomes have been stagnant for decades is, after all, one of the primary problems we are trying to address. I don't see anything especially dystopian in offering some relatively unproductive people a minimal income as an incentive to leave the workforce, as long as the result is more opportunity and higher incomes for those who do want to work hard and advance their situation.

While our value system is geared toward celebrating production, it's important to keep in mind that consumption is also an essential

---

* Obviously, I'm leaving aside those people who might choose to drop out of the workforce (at least temporarily) for reasons we would likely consider more legitimate, such as caring for children or other family members. For some families, for example, a basic income might turn out to be a partial solution to the looming elder-care problem.

economic function. The person who takes the income and drops out will become a paying customer for the hardworking entrepreneur who sets up a small business in the same neighborhood. And that businessperson will, of course, receive the same basic income.

A final point is that most policy errors in implementing a guaranteed income ought to eventually be self-correcting. If the income were initially too generous and thereby resulted in a strong disincentive to work, then one of two things would happen. Either automation technology would be sufficiently advanced to pick up the slack in production (in which case there would be no problem), or there would be a labor shortage and a burst of inflation. A general increase in prices would devalue the basic income and re-create the incentive to supplement it with work. Unless policy makers did something truly misguided—like, for example, building an automatic cost-of-living increase into the income scheme—any inflation would probably be short-lived, and then the economy would find a new equilibrium.

Beyond the political challenges and risks associated with a general disincentive to work, there is also the question of the impact a basic income might have on housing costs in high-rent areas. Imagine giving every resident of a city like New York, San Francisco, or London an extra thousand dollars per month. There are probably good reasons to expect that a very large fraction of that increase—perhaps nearly all of it—would eventually end up in the pockets of landlords as residents compete for scarce housing. There are no easy solutions to this problem. Rent control is one possibility, but it comes with lots of documented downsides. Many economists have called for relaxing zoning restrictions so that denser housing can be built, but this is sure to be opposed by existing residents.

There is a counteracting force, however. A guaranteed income, unlike a job, would be mobile. Some people would be very likely to take their income and move away from expensive areas in search of a lower cost of living. There might be an influx of new residents into declining cities like Detroit. Others would choose to leave cities altogether. A basic income program might help revitalize many of the small towns and rural areas that are losing population because jobs have evaporated. Indeed, I think the potentially positive economic impact on rural areas might be one factor that could help make a guaranteed income policy attractive to conservatives in the United States.

Immigration policy is another area that would obviously need to be adjusted in the wake of the implementation of a guaranteed income. It seems likely that immigration as well as any subsequent path to citizenship and eligibility for the income would have to be restricted, or perhaps a significant waiting period would need to be imposed for new citizens. All of this would, of course, add even more complexity and uncertainty to a political issue that is already intensely polarizing.

Paying for a Basic Income

If the United States were to give every adult between the ages of twenty-one and sixty-five, as well as those over sixty-five who are not receiving Social Security or a pension, an unconditional annual income of $10,000, the total cost would be somewhere in the vicinity of $2 trillion.[17] This amount would be reduced somewhat by limiting eligibility for the basic income to citizens and perhaps by means-testing it against earned income beyond a certain point. (As I suggested earlier, it would be very important to phase the guaranteed income out only at a fairly high level in order to avoid a poverty trap scenario.) The total cost would then be offset by reducing or eliminating numerous federal and state anti-poverty programs, including food stamps, welfare, housing assistance, and the Earned Income Tax Credit. (The EITC is discussed in further detail below.) These programs add up to as much as $1 trillion per year.

In other words, a $10,000 annual basic income would probably require around $1 trillion in new revenue, or perhaps significantly less if we instead chose some type of guaranteed minimum income. That number would be further reduced, however, by increased tax revenues resulting from the plan. The basic income itself would be taxable, and it would likely push many households out of Mitt Romney's infamous "47 percent" (the fraction of the population who currently pay no federal income tax). Most lower-income households would spend most of their basic income, and that would result directly in more taxable economic activity. Given that advancing technology is likely to drive us toward higher levels of inequality while undermining broad-based consumption, a guaranteed income might well result in a significantly higher rate of economic growth over the long run—and that, of course, would mean much higher tax revenue. And since a basic income would keep a consistent flow of purchasing power streaming to consumers, it would act as a powerful economic stabilizer, allowing the economy to avoid some of the costs associated with deep recessions. All these effects are, of course, difficult to quantify, but I think there is a strong argument that a basic income would, at least to some extent, pay for itself. Furthermore, the economic gains from its implementation would increase over time as technology advances and the economy becomes ever more capital-intensive.

It goes without saying that raising sufficient revenue would be an enormous challenge in today's political environment, given that nearly all American politicians are terrified to even utter the word "tax" unless it is followed immediately by the word "cut." The most feasible approach might be to use a variety of different taxes to raise the necessary revenue. One obvious candidate would be a carbon tax, which could raise as much as $100 billion per year while helping to reduce greenhouse gas emissions. There have already been proposals for a revenue-neutral carbon tax with a rebate for every household, and this might serve as a starting point for a basic income. Another option is a value-added tax. The United States is the only advanced nation that does not currently rely on such a tax—essentially a type of consumption tax that gets tacked on at every step in the production process. A VAT is passed on to consumers as part of the final price charged for products and services and is generally considered to be a very efficient way to raise tax revenue. There are numerous other possibilities, including higher corporate taxes (or the elimination of tax avoidance schemes), some type of national land tax, higher capital gains taxes, and a financial transaction tax.

In seems inevitable that personal income taxes would also have to increase, and one of the best ways to do this is to make the system more progressive. One of the implications of increasing inequality is that ever more taxable income is rising to the very top. Our taxation scheme should be restructured to mirror the income distribution. Rather than simply raising taxes across the board or on the highest existing tax bracket, a better strategy would be to introduce several new higher tax brackets designed to capture more revenue from those taxpayers with very high incomes—perhaps a million or more dollars per year.

Everyone a Capitalist

While I believe that some form of guaranteed income is probably the best overall solution to the rise of automation technology, there are certainly other viable ideas. One of the most common proposals is to focus on wealth, rather than income. In a future world where nearly all the income is captured by capital, and human labor is worth very little, why not simply make sure that everyone owns enough capital to be economically secure?

Most of these proposals involve strategies like somehow increasing employee stock ownership in businesses or simply giving everyone a substantial balance in a mutual fund. In an article for *The Atlantic,* economist Noah Smith suggests that the government could give everyone "an endowment of capital" by purchasing a "diversified portfolio of equity" for every citizen when he or she turns eighteen. A rash decision to "cash out, and party" would be "prevented with some fairly light paternalism, like temporary 'lock-up' provisions."[18]

The problem with this is that "light paternalism" might not be enough. Imagine a future in which your ability to survive economically is determined almost exclusively by what you own; your labor is worth little or nothing. In that world, there would be no more stories about the person who lost it all and then worked his or her way back to the top. If you make a bad investment or get ripped off by a Bernie Madoff type, then the error might well be unrecoverable. If individuals are ultimately given control over their capital, then it's inevitable that this scenario would play out for some unlucky people. What would we do for individuals and families who found themselves in this kind of situation? Would they be "too big to fail"? If so, there would be a clear moral hazard problem: people might see little downside in taking excessive risks. If not, we'd have people in genuinely dire situations with little or no hope of escape.

The vast majority of people would, of course, act responsibly in the face of this kind of risk. But that might result in its own problems. If loss of your capital meant destitution for you and your children, would you be willing to invest a chunk of it in a new business venture? Experience with 401k retirement plans has shown that many people elect to invest too little in the stock market and too much in lower-return investments they perceive as safe. In a world where capital is everything, that preference might well be amplified. There could be huge demand for safe assets, and as a result the returns on those assets would be very low. In other words, a solution based on giving people wealth might result in something quite different from the Peltzman effect I suggested we might see with a guaranteed income. Excessive risk aversion could lead to less entrepreneurship, lower incomes, and less vibrant market demand.*

Yet another problem, of course, is paying for these equity endowments. My guess is that redistribution of vast amounts of capital

* Some economists, most notably former US Treasury secretary Larry Summers, have suggested that the economy is currently trapped in "secular stagnation"—a situation where interest rates are near zero, the economy is operating below its potential, and there is too little investment in more productive opportunities. I think a future where everyone is dependent almost entirely on his or her mutual fund balance for economic survival might well result in a similar outcome.

would prove even more politically toxic than would be the case for income. One possible mechanism for prying wealth away from its current owners was proposed by Thomas Piketty in his book *Capital in the Twenty-First Century:* a global tax on wealth. Such a tax would require cooperation between nations in order to avoid massive capital flight into lower-tax jurisdictions. Nearly everyone (including Piketty) agrees that this would be impractical for the foreseeable future.

Piketty's book, which was deluged with attention in 2014, argues that future decades are likely to be marked by an inevitable progression toward increased inequality of both income and wealth. Piketty approaches the issue of inequality purely from the perspective of a historical analysis of economic data. His central thesis is that the return on capital is usually greater than the overall rate of economic growth, so that capital ownership inevitably becomes a larger slice of the economic pie over time. He shows surprisingly little interest in the trends we've focused on here; indeed, the word "robot" appears on only one of his book's nearly seven hundred pages. If Piketty's theory is correct—and it has been subject to a great deal of debate—then I think advancing technology is likely to greatly amplify his conclusions, quite possibly producing even higher levels of future inequality than his model predicts.

It's possible that as the issue of inequality, and especially its impact on the political process in the United States, gains ever more visibility with the public, the kind of wealth tax that Piketty advocates might someday become viable. If so, I would argue that rather than portioning out redistributed capital to individuals, it would be better to set up a centrally managed sovereign wealth fund (similar to the Alaska fund) and then use the resulting returns to help fund a basic income.

Near-Term Policies

While the establishment of a guaranteed income will probably remain politically unfeasible for the foreseeable future, there are a number of other things that might prove helpful in the nearer term.

Many of these ideas are really generic economic policies geared toward enabling a more robust recovery from the Great Recession. In other words, they are things we ought to be doing in any case, independently of any concern about the impact of robots or automation on jobs.

Foremost among these policies is the critical need for the United States to invest in public infrastructure. There is an enormous pent-up requirement to repair and refurbish things like roads, bridges, schools, and airports. This maintenance will have to be performed eventually; there is no getting around it, and the longer we wait the more it will ultimately cost. The federal government can currently borrow money at interest rates remarkably close to zero, while unemployment among construction workers remains at double-digit rates. Our failure to take advantage of this opportunity and make the necessary investments while the cost is low is likely to someday be judged to be economic malpractice of the highest order.

While I'm skeptical that policies geared toward more education and vocational training will offer a long-term, systemic solution to the problem of technological unemployment, there are certainly many things we can and should be doing to improve the more immediate prospects for students and workers. We probably can't change the reality that there will be only a limited number of jobs available at the top of the skills pyramid. However, we certainly can address the issue of workers who don't have the necessary skills for the opportunities that do exist. In particular, there is an obvious need for more investment in community colleges. Some professions with low unemployment rates, especially health care–related fields like nursing, are currently subject to significant educational bottlenecks; there is overwhelming demand for training, but students are unable to get into classes that are filled beyond capacity. In general, community colleges represent one of our most important resources for enabling workers to navigate an increasingly dynamic job market. Given that jobs—and entire occupations—may be poised to evaporate at an accelerating pace, we should do everything possible to make opportunities for retraining available. Expanding access to relatively inexpensive community colleges, while doing more to rein in predatory for-profit schools that have been set up primarily to harvest financial aid dollars, would result in improved prospects for a great many people. As we saw in Chapter 5, MOOCs and other innovations in online education may also eventually have a meaningful impact on vocational training opportunities.

Another important proposal centers on expansion of the Earned Income Tax Credit, a subsidy paid to low-income workers in the United States. The EITC is currently subject to two primary limitations. First, the unemployed are not eligible; in order to ensure an incentive to work, the benefit is paid only to people who have earned income. Second, the program is primarily configured as a form of child support. A single parent with three or more children could get a maximum of about $6,000 per year in 2013, while a childless worker could receive only $487—or about $40 per month. The Obama administration has already proposed to expand eligibility for workers without children, although the maximum benefit would still only be about $1,000 per year. Transforming the EITC into a viable longer-term solution would require extending eligibility to those who are unable to find jobs—and that, of course, would amount to converting the program into a guaranteed income. The near-term prospects for expanding the EITC in any way seem bleak, as Republicans in Congress have expressed a desire to actually cut the program.

If you accept the argument that our economy is likely to become ever less labor-intensive over time, then it follows logically that we ought to shift our taxation scheme away from labor and toward capital. Currently, major programs that support the elderly, for example, are funded largely by payroll taxes that fall on both workers and employers. Taxing work in this way allows those businesses that are highly capital- or technology-intensive to, in a sense, free-ride—reaping the benefits of our markets and institutions while escaping their obligation to contribute to the support of programs that are critical to society as a whole. As the taxation burden falls disproportionately on more labor-intensive industries and businesses, it will further increase the incentive to shift away from human labor and toward automation whenever possible. Eventually, the entire system could well become unsustainable. Instead, we ought to transition to a form of taxation that asks more from those businesses that rely heavily on technology and employ relatively few workers. We eventually will have to move away from the idea that workers support retirees and pay for social programs, and instead adopt the premise that our overall economy supports these things. Economic growth, after all, has significantly outpaced the rate at which new jobs have been created and wages have been rising.

If these proposals strike you as overly ambitious, then there remains at least one policy prescription that ought to be straightforward. Given the trends we've reviewed in these pages, it seems evident that we should not now be setting out to dismantle the social safety net we currently have in place. If there is, in fact, any good time to slash the programs that the most vulnerable segments of our population rely on—without also putting in place a viable alternate solution—then, surely, *this is not that time*.

THE POLITICAL ENVIRONMENT in the United States has become so toxic and divisive that agreement on even the most conventional economic policies seems virtually impossible. Given this, it's easy to dismiss any talk of more radical interventions like a guaranteed income as completely pointless. There is an understandable temptation to focus exclusively on smaller, possibly more feasible, policies that might nibble at the margins of our problems, while leaving any discussion of the larger challenges for some indeterminate point in the future.

This is dangerous because we are now so far along on the arc of information technology's progress. We are getting onto the steep part of the exponential curve. Things will move faster, and the future may arrive long before we are ready.

The decades-long struggle to adopt universal health coverage in the United States probably offers a pretty good preview of the staggering challenge we will face in attempting to bring about any kind of whole-scale economic reform. Nearly eighty years passed from the time Franklin Roosevelt first proposed a national health care system until the passage of the Affordable Care Act. In the case of health care, of course, America had as working examples the long-established systems of every other advanced nation in the world. But there are no examples of a working guaranteed income—or, for that matter, any other policy designed to adapt to the implications of future technology. We will have to make it up as we go along. Given this, it is surely not too soon to begin a meaningful discussion.

That discussion will have to delve into our fundamental assumptions about the role of labor in our economy and the way people respond to incentives. Everyone agrees that incentives are important, but there are good reasons to believe that our economic incentives could safely be moderated somewhat. This is true at both ends of the income spectrum. The premise that even modestly higher marginal tax rates on top incomes will somehow destroy the impetus for entrepreneurship and investment is simply unsupportable. The fact that both Apple and Microsoft were founded in the mid-1970s—a period when the top tax bracket stood at 70 percent—offers pretty good evidence that entrepreneurs don't spend a lot of time worrying about top tax rates. Likewise, at the bottom, the motivation to work certainly matters, but in a country as wealthy as the United States, perhaps that incentive does not need to be so extreme as to elicit the specters of homelessness and destitution. Our fear that we will end up with too many people riding in the economic wagon, and too few pulling it, ought to be reassessed as machines prove increasingly capable of doing the pulling.

In May 2014, payroll employment in the United States finally returned to its pre-recession peak, bringing to an end an epic jobless recovery that spanned more than six years. Even as total employment recovered, however, there was general agreement that the quality of those jobs was significantly diminished. The crisis had wiped out millions of middle-class jobs, while the positions created over the course of the recovery were disproportionately in low-wage service industries. A great many were in fast food and retail occupations—areas that, as we have seen, seem very likely to eventually be impacted by advances in robotics and self-service automation. Both long-term unemployment and the number of people unable to find full-time work remain at elevated levels.

Lurking behind the headline employment figure was another number that carried with it an ominous warning for the future. In the years since the onset of the financial crisis, the population of working-age adults in the United States had increased by about 15 million people.[19] For all those millions of entrants into the workforce, the economy had created no new opportunities at all. As John Kennedy said, "To even stand still we have to move very fast." That was possible in 1963. In our time, it may ultimately prove unachievable.

CONCLUSION

In the same month that the total number of jobs in the United States finally returned to pre-crisis levels, the US government released two reports that offer some perspective on the magnitude and complexity of the challenges we are likely to face in the coming decades. The first, which went almost completely unnoticed, was a brief analysis published by the Bureau of Labor Statistics. The report looked at how the total amount of work performed in the US private sector had changed over the course of fifteen years. Rather than simply counting jobs, the BLS delved into the actual number of hours worked.

In 1998, workers in the US business sector put in a total of 194 billion hours of labor. A decade and a half later, in 2013, the value of the goods and services produced by American businesses had grown by about $3.5 trillion after adjusting for inflation—a 42 percent increase in output. The total amount of human labor required to accomplish that was . . . 194 billion hours. Shawn Sprague, the BLS economist who prepared the report, noted that "this means that there was ultimately *no growth at all* in the number of hours worked over this 15-year period, despite the fact that the US population gained over 40 million people during that time, and despite the fact that there were thousands of new businesses established during that time."[1]

News of the second report, which was released on May 6, 2014, was splashed across the front page of the *New York Times*. "The

National Climate Assessment," a major interagency project supervised by a sixty-member panel that included representatives from the oil industry, declared that "climate change, once considered an issue for a distant future, has moved firmly into the present."[2] The report noted that "summers are longer and hotter, and extended periods of unusual heat last longer than any living American has ever experienced." The United States has already seen a dramatic increase in the frequency of torrential rains, often leading to flooding and widespread damage. The report projected a sea-level rise of between one and four feet by 2100 and noted that already "residents of some coastal cities see their streets flood more regularly during storms and high tides." The market economy has begun to adjust to the reality of climate change; flood insurance is increasing in cost, or even becoming completely unavailable, in vulnerable areas.

Among techno-optimists, there is a tendency to discount concerns about climate change and environmental impact. Technology is viewed along a single dimension: it is a universally positive force whose exponential progress will almost surely rescue us from any dangers that lie ahead. Abundant clean energy will power our economy long before we expect it, and innovations in areas like the desalination of ocean water and more efficient recycling will arrive in time to head off any dramatically negative consequences. Some level of optimism is certainly justified. Solar power, in particular, has recently been subject to a Moore's Law–like trend that is rapidly bringing costs down. Global installed photo-voltaic capacity has been doubling roughly every two and a half years.[3] The most extreme optimists think we will be able to get *all* our power from solar by the early 2030s.[4] Still, significant challenges remain; one problem is that, while the cost of solar panels themselves has been declining rapidly, other important costs—such as those of peripheral equipment and installation—have not, so far, been subject to the same rate of progress.

A more realistic view suggests that we will need to rely on a combination of both innovation and regulation if we are going to successfully mitigate and adapt to climate change. The story of the future is not going to be about a simple contest between technology and environmental impact. It will be far more complicated than that. As we have seen, advancing information technology has a dark side of its own, and if it results in widespread unemployment or threatens the economic security of a large fraction of our population, the dangers posed by climate change will become politically even more difficult to address.

A 2013 survey by researchers at Yale and George Mason Universities found that about 63 percent of Americans believe climate change is happening, and that just over half are at least somewhat worried about its future implications.[5] A more recent survey by Gallup probably puts things in better perspective, however.[6] On a list of fifteen major concerns, climate change came in at number fourteen. First on the list was the economy, and for the vast majority of average people "the economy," of course, really amounts to jobs and the wages they pay.

History shows clearly that when jobs are scarce, the fear of even more unemployment becomes a powerful tool in the hands of politicians and special interests who oppose action on the environment. This has been the case, for example, in those states where coal mining has historically been an important source of jobs, despite the fact that employment in the mining industry has been decimated not by environmental regulation but by mechanization. Corporations with even small numbers of jobs to offer routinely play states and cities against each other, seeking lower taxes, government subsidies, and freedom from regulation.

Beyond the United States and other advanced countries, the situation may be far more dangerous. As we've seen, factory jobs are disappearing across the globe at a rapid clip. Labor-intensive manufacturing as a path to prosperity may begin to evaporate for many developing nations even as more efficient farming techniques inevitably push people away from agricultural lifestyles. Many of these countries will see far more severe impacts from climate change and are already subject to significant environmental degradation. In the worst-case scenario, a combination of widespread economic insecurity, drought, and rising food prices could eventually lead to social and political instability.

The greatest risk is that we could face a "perfect storm"—a situation where technological unemployment and environmental impact unfold roughly in parallel, reinforcing and perhaps even amplifying each other. If, however, we can fully leverage advancing technology as a solution—while recognizing and adapting to its implications for employment and the distribution of income—then the outcome is likely to be far more optimistic. Negotiating a path through these entangled forces and crafting a future that offers broad-based security and prosperity may prove to be the greatest challenge for our time.

ACKNOWLEDGMENTS

First and foremost, I would like to thank the entire team at Basic Books—and especially my extraordinary editor, T. J. Kelleher—for working with me to make this book a reality. My agent, Don Fehr of Trident Media, was instrumental in helping this project find its proper home at Basic Books.

I am also extremely grateful to the many readers of my earlier book, *The Lights in the Tunnel,* who wrote to me with suggestions and criticisms, as well as examples demonstrating how the relentless trend toward automation is unfolding in the real world. Many of these ideas and discussions helped to refine my thinking as I approached this book. In particular, I thank Abhas Gupta of Mohr Davidow Ventures, who pointed me to some of the specific examples cited in these pages and also offered many valuable suggestions after reading an early draft of the book.

Many of the graphs and charts in this book were generated using data from the excellent Federal Reserve Economic Data (FRED) system provided by the Federal Reserve Bank of St. Louis. The specific data series I used are given in the notes. I encourage any interested readers to visit the FRED website and experiment with this remarkable resource. I also thank Lawrence Mishel of the Economic Policy Institute for allowing me to reproduce his analysis showing the dramatic divergence of growth in productivity and compensation in the United States, and Simon Colton for providing an illustration created by his artistic AI application, "The Painting Fool."

Finally, I thank my family, and especially my lovely wife, Xiaoxiao Zhao, for their support and patience during the long process (and many late nights) that ultimately led to this book.

NOTES

INTRODUCTION

1. Average wages for production or nonsupervisory workers: *The Economic Report of the President, 2013,* Table B-47, http://www.whitehouse.gov/sites/default/files/docs/erp2013/full_2013_economic_report_of_the_president.pdf. The table shows peak weekly wages of about $341 in 1973 and $295 in December 2012, measured in 1984 dollars. Productivity: Data Source: FRED, Federal Reserve Economic Data, Federal Reserve Bank of St. Louis: Nonfarm Business Sector: Real Output Per Hour of All Persons, Index 2009=100, Seasonally Adjusted [OPHNFB]; US Department of Labor: Bureau of Labor Statistics; https://research.stlouisfed.org/fred2/series/OPHNFB/; accessed April 29, 2014.

2. Neil Irwin, "Aughts Were a Lost Decade for U.S. Economy, Workers," *Washington Post,* January 2, 2010, http://www.washingtonpost.com/wp-dyn/content/article/2010/01/01/AR2010010101196.html.

3. Ibid.

CHAPTER 1

1. John Markoff, "Skilled Work, Without the Worker," *New York Times,* August 18, 2012, http://www.nytimes.com/2012/08/19/business/new-wave-of-adept-robots-is-changing-global-industry.html.

2. Damon Lavrinc, "Peek Inside Tesla's Robotic Factory," *Wired.com,* July 16, 2013, http://www.wired.com/autopia/2013/07/tesla-plant-video/.

3. International Federation of Robotics website, Industrial Robot Statistics 2013, http://www.ifr.org/industrial-robots/statistics/.

4. Jason Tanz, "Kinect Hackers Are Changing the Future of Robotics," *Wired Magazine,* July 2011, http://www.wired.com/magazine/2011/06/mf_kinect/.

5. Esther Shein, "Businesses Adopting Robots for New Tasks," *Computerworld,* August 1, 2013, http://www.computerworld.com/s/article/9241118/Businesses_adopting_robots_for_new_tasks.

6. Stephanie Clifford, "U.S. Textile Plants Return, with Floors Largely Empty of People," *New York Times,* September 12, 2013, http://www.nytimes.com/2013/09/20/business/us-textile-factories-return.html.

7. Ibid.

8. On increasing worker pay in China and Boston Consulting Group Survey, see "Coming Home," *The Economist,* January 19, 2013, http://www.economist.com/news/special-report/21569570-growing-number-american-companies-are-moving-their-manufacturing-back-united.

9. Caroline Baum, "So Who's Stealing China's Manufacturing Jobs?," *Bloomberg News,* October 14, 2003, http://www.bloomberg.com/apps/news?pid=newsarchive&sid=aRI4bAft7Xw4.

10. Paul Mozur and Eva Dou, "Robots May Revolutionize China's Electronics Manufacturing," *Wall Street Journal,* September 24, 2013, http://online.wsj.com/news/articles/SB10001424052702303759604579093122607195610.

11. For more on China's artificially low cost of capital, see Michael Pettis, *Avoiding the Fall: China's Economic Restructuring* (Washington, DC: Carnegie Endowment for International Peace, 2013).

12. Barney Jopson, "Nike to Tackle Rising Asian Labour Costs," *Financial Times,* June 27, 2013, http://www.ft.com/intl/cms/s/0/277197a6-df6a-11e2–881f-00144feab7de.html.

13. Momentum Machines co-founder Alexandros Vardakostas, as quoted in Wade Roush, "Hamburgers, Coffee, Guitars, and Cars: A Report from Lemnos Labs," *Xconomy.com,* June 12, 2012, http://www.xconomy.com/san-francisco/2012/06/12/hamburgers-coffee-guitars-and-cars-a-report-from-lemnos-labs/.

14. Momentum Machines website, http://momentummachines.com; David Szondy, "Hamburger-Making Machine Churns Out Custom Burgers at Industrial Speeds," *Gizmag.com,* November 25, 2012, http://www.gizmag.com/hamburger-machine/25159/.

15. McDonald's corporate website, http://www.aboutmcdonalds.com/mcd/our_company.html.

16. US Department of Labor, Bureau of Labor Statistics, News Release, December 19, 2013, USDL-13–2393, Employment Projections—2012–2022, Table 8, http://www.bls.gov/news.release/pdf/ecopro.pdf.

17. Alana Semuels, "National Fast-Food Wage Protests Kick Off in New York," *Los Angeles Times,* August 29, 2013, http://articles.latimes.com/2013

/aug/29/business/la-fi-mo-fast-food-protests-20130829.

18. Schuyler Velasco, "McDonald's Helpline to Employee: Go on Food Stamps," *Christian Science Monitor,* October 24, 2013, http://www.csmonitor.com/Business/2013/1024/McDonald-s-helpline-to-employee-Go-on-food-stamps.

19. Sylvia Allegretto, Marc Doussard, Dave Graham-Squire, Ken Jacobs, Dan Thompson, and Jeremy Thompson, "Fast Food, Poverty Wages: The Public Cost of Low-Wage Jobs in the Fast-Food Industry," UC Berkeley Labor Center, October 15, 2013, http://laborcenter.berkeley.edu/publiccosts/fast_food_poverty_wages.pdf.

20. Hiroko Tabuchi, "For Sushi Chain, Conveyor Belts Carry Profit," *New York Times,* December 30, 2010, http://www.nytimes.com/2010/12/31/business/global/31sushi.html.

21. Stuart Sumner, "McDonald's to Implement Touch-Screen Ordering," *Computing,* May 18, 2011, http://www.computing.co.uk/ctg/news/2072026/mcdonalds-implement-touch-screen.

22. US Department of Labor, Bureau of Labor Statistics, *Occupational Outlook Handbook,* March 29, 2012, http://www.bls.gov/ooh/About/Projections-Overview.htm.

23. Ned Smith, "Picky Robots Grease the Wheels of e-Commerce," *Business News Daily,* June 2, 2011, http://www.businessnewsdaily.com/1038-robots-streamline-order-fulfillment-e-commerce-pick-pack-and-ship-warehouse-operations.html.

24. Greg Bensinger, "Before Amazon's Drones Come the Robots," *Wall Street Journal,* December 8, 2013, http://online.wsj.com/news/articles/SB10001424052702303330204579246012421712386.

25. Bob Trebilcock, "Automation: Kroger Changes the Distribution Game," *Modern Materials Handling,* June 4, 2011, http://www.mmh.com/article/automation_kroger_changes_the_game.

26. Alana Semuels, "Retail Jobs Are Disappearing as Shoppers Adjust to Self-Service," *Los Angeles Times,* March 4, 2011, http://articles.latimes.com/2011/mar/04/business/la-fi-robot-retail-20110304.

27. Redbox corporate blog, "A Day in the Life of a Redbox Ninja," April 12, 2010, http://blog.redbox.com/2010/04/a-day-in-the-life-of-a-redbox-ninja.html.

28. Redbox corporate website, http://www.redbox.com/career-technology.

29. Meghan Morris, "It's Curtains for Blockbuster's Remaining U.S. Stores," *Crain's Chicago Business,* November 6, 2013, http://www.chicagobusiness.com/article/20131106/NEWS07/131109882/its-curtains-for-blockbusters-remaining-u-s-stores.

30. Alorie Gilbert, "Why So Nervous About Robots, Wal-Mart?," *CNET News,* July 8, 2005, http://news.cnet.com/8301–10784_3–5779674–7.html.

31. Jessica Wohl, "Walmart Tests iPhone App Checkout Feature," Reuters, September 6, 2012, http://www.reuters.com/article/2012/09/06/us-walmart-iphones-checkout-idUSBRE8851DP20120906.

32. Brian Sumers, "New LAX Car Rental Company Offers Only Audi A4s—and No Clerks," *Daily Breeze,* October 6, 2013, http://www.dailybreeze.com/general-news/20131006/new-lax-car-rental-company-offers-only-audi-a4s-x2014-and-no-clerks.

33. Vision Robotics corporate website, http://visionrobotics.com.

34. Harvest Automation corporate website, http://www.harvestai.com/agricultural-robots-manual-labor.php.

35. Peter Murray, "Automation Reaches French Vineyards with a Vine-Pruning Robot," *SingularityHub,* November 26, 2012, http://singularityhub.com/2012/11/26/automation-reaches-french-vineyards-with-a-vine-pruning-robot.

36. "Latest Robot Can Pick Strawberry Fields Forever," *Japan Times,* September 26, 2013, http://www.japantimes.co.jp/news/2013/09/26/business/latest-robot-can-pick-strawberry-fields-forever.

37. Australian Centre for Field Robotics website, http://sydney.edu.au/engineering/research/robotics/agricultural.shtml.

38. Emily Sohn, "Robots on the Farm," *Discovery News,* April 12, 2011, http://news.discovery.com/tech/robotics/robots-farming-agriculture-110412.htm.

39. Alana Semuels, "Automation Is Increasingly Reducing U.S. Workforces," *Los Angeles Times,* December 17, 2010, http://articles.latimes.com/2010/dec/17/business/la-fi-no-help-wanted-20101217.

Chapter 2

1. On Martin Luther King Jr.'s final sermon and memorial service at Washington National Cathedral, see Ben A. Franklins, "Dr. King Hints He'd Cancel March If Aid Is Offered," *New York Times,* April 1, 1968, and Nan Robertson, "Johnson Leads U.S. in Mourning: 4,000 Attend Service at Cathedral in Washington," *New York Times,* April 6, 1968.

2. The full text of MLK's "Remaining Awake Through a Great Revolution" sermon is available at http://mlk-kpp01.stanford.edu/index.php/kingpapers/article/remaining_awake_through_a_great_revolution/.

3. For the text of the Triple Revolution report and a list of signatories, see http://www.educationanddemocracy.org/FSCfiles/C_CC2a_TripleRevolution.htm. Scanned images of the original document and accompanying letter to President Johnson are available at http://osulibrary.oregonstate.edu/specialcollections/coll/pauling/peace/papers/1964p.7–04.html.

4. John D. Pomfret, "Guaranteed Income Asked for All, Employed or Not," *New York Times,* March 22, 1964. For other media coverage of the

Triple Revolution report, see Brian Steensland, *The Failed Welfare Revolution: America's Struggle over Guaranteed Income Policy* (Princeton: Princeton University Press, 2011), pp. 43–44.

5. Norbert Wiener's article on automation is discussed and quoted extensively in John Markoff, "In 1949, He Imagined an Age of Robots," *New York Times,* May 20, 2013.

6. From a letter to Robert Weide dated January 12, 1983, cited in Dan Wakefield, ed., *Kurt Vonnegut Letters* (New York: Delacorte Press, 2012), p. 293.

7. For the text of Lyndon B. Johnson's "Remarks Upon Signing Bill Creating the National Commission on Technology, Automation, and Economic Progress," August 19, 1964, see Gerhard Peters and John T. Woolley, *The American Presidency Project,* http://www.presidency.ucsb.edu/ws/?pid=26449.

8. The National Commission on Technology, Automation, and Economic Progress reports can be found online at http://catalog.hathitrust.org/Record/009143593, http://catalog.hathitrust.org/Record/007424268, and http://www.rand.org/content/dam/rand/pubs/papers/2013/P3478.pdf.

9. For information on unemployment rates in the 1950s and '60s, see "A Brief History of US Unemployment" at the *Washington Post* website, http://www.washingtonpost.com/wp-srv/special/business/us-unemployment-rate-history/.

10. For a vivid description of the design and operation of the first digital computers and the teams that built them, see George Dyson, *Turing's Cathedral: The Origins of the Digital Universe* (New York: Vintage, 2012).

11. For a listing of the average wages for production or nonsupervisory workers, see Table B-47 in *The Economic Report of the President, 2013,* http://www.whitehouse.gov/sites/default/files/docs/erp2013/full_2013_economic_report_of_the_president.pdf. As noted in the Introduction, the table shows peak weekly wages of about $341 in 1973 and $295 in December 2012, measured in 1984 dollars. I have adjusted these to 2013 dollars using the Bureau of Labor Statistics' inflation calculator at http://www.bls.gov/data/inflation_calculator.htm.

12. On median household incomes versus per capita GDP, see Tyler Cowen, *The Great Stagnation: How America Ate All the Low-Hanging Fruit of Modern History, Got Sick, and Will (Eventually) Feel Better* (New York: Dutton, 2011), p. 15, and Lane Kenworthy, "Slow Income Growth for Middle America," September 3, 2008, http://lanekenworthy.net/2008/09/03/slow-income-growth-for-middle-america/. I have adjusted the figures to reflect 2013 dollars.

13. Lawrence Mishel, "The Wedges Between Productivity and Median Compensation Growth," Economic Policy Institute, April 26, 2012, http://www.epi.org/publication/ib330-productivity-vs-compensation/.

14. "The Compensation-Productivity Gap," US Bureau of Labor Statistics website, February 24, 2011, http://www.bls.gov/opub/ted/2011/ted_20110224.htm.

15. John B. Taylor and Akila Weerapana, *Principles of Economics* (Mason, OH: Cengage Learning, 2012), p. 344. In particular, see the bar chart and commentary in the left margin. Taylor is a very highly regarded economist, known especially for the "Taylor Rule," a monetary policy guideline used by central banks (including the Federal Reserve) to set interest rates.

16. Robert H. Frank and Ben S. Bernanke, *Principles of Economics*, 3rd ed. (New York: McGraw Hill/Irwin, 2007), pp. 596–597.

17. John Maynard Keynes, as quoted in David Hackett Fischer, *The Great Wave: Price Revolutions and the Rhythm of History* (New York: Oxford University Press, 1996), p. 294.

18. Labor Share Graph, Data Source: FRED, Federal Reserve Economic Data, Federal Reserve Bank of St. Louis: Nonfarm Business Sector: Labor Share, Index 2009=100, Seasonally Adjusted [PRS85006173]; US Department of Labor: Bureau of Labor Statistics; https://research.stlouisfed.org/fred2/series/PRS85006173; accessed April 29, 2014. The vertical scale is an index with 100 set to 2009. The labor share percentages shown on the graph (65% and 58%) were added for clarity. See also: Margaret Jacobson and Filippo Occhino, "Behind the Decline in Labor's Share of Income," Federal Reserve Bank of Cleveland, February 3, 2012 (http://www.clevelandfed.org/research/trends/2012/0212/01gropro.cfm).

19. Scott Thurm, "For Big Companies, Life Is Good," *Wall Street Journal*, April 9, 2012, http://online.wsj.com/article/SB1000142405270230381540457731660464739018.html.

20. Ibid.

21. Corporate Profits / GDP graph: Data Source: FRED, Federal Reserve Economic Data, Federal Reserve Bank of St. Louis: Corporate Profits After Tax (without IVA and CCAdj), Billions of Dollars, Seasonally Adjusted Annual Rate [CP]; Gross Domestic Product, Billions of Dollars, Seasonally Adjusted Annual Rate [GDP]; http://research.stlouisfed.org/fred2/graph/?id=CP; accessed April 29, 2014.

22. Loukas Karabarbounis and Brent Neiman, "The Global Decline of the Labor Share," National Bureau of Economic Research, Working Paper No. 19136, issued in June 2013, http://www.nber.org/papers/w19136.pdf; see also http://faculty.chicagobooth.edu/loukas.karabarbounis/research/labor_share.pdf.

23. Ibid., p. 1.

24. Ibid.

25. Labor Force Participation Rate Graph, Data Source: FRED, Federal Reserve Economic Data, Federal Reserve Bank of St. Louis: Civilian Labor Force Participation Rate, Percent, Seasonally Adjusted [CIVPART]; http://research.stlouisfed.org/fred2/graph/?id=CIVPART; accessed April 29, 2014.

26. Graphs showing the participation rates for men and women can be found at the Federal Reserve Economic Data website; see http://research.stlouisfed.org/fred2/series/LNS11300001 and http://research.stlouisfed.org/fred2/series/LNS11300002, respectively.

27. A graph of the labor force participation rate for adults twenty-five to fifty-four years of age can be found at http://research.stlouisfed.org/fred2/graph/?g=l6S.

28. On the greatly increased number of disability applications, see Willem Van Zandweghe, "Interpreting the Recent Decline in Labor Force Participation," *Economic Review—First Quarter 2012,* Federal Reserve Bank of Kansas City, p. 29, http://www.kc.frb.org/publicat/econrev/pdf/12q1VanZandweghe.pdf.

29. Data Source: FRED, Federal Reserve Economic Data, Federal Reserve Bank of St. Louis: All Employees: Total Nonfarm, Thousands of Persons, Seasonally Adjusted [PAYEMS]; US Department of Labor: Bureau of Labor Statistics; https://research.stlouisfed.org/fred2/series/PAYEMS/; accessed June 10, 2014.

30. On the number of new jobs needed to keep up with population growth, see Catherine Rampell, "How Many Jobs Should We Be Adding Each Month?," *New York Times* (Economix blog), May 6, 2011, http://economix.blogs.nytimes.com/2011/05/06/how-many-jobs-should-we-be-adding-each-month/.

31. Murat Tasci, "Are Jobless Recoveries the New Norm?," Federal Reserve Bank of Cleveland, Research Commentary, March 22, 2010, http://www.clevelandfed.org/research/commentary/2010/2010–1.cfm.

32. Center on Budget and Policy Priorities, "Chart Book: The Legacy of the Great Recession," September 6, 2013, http://www.cbpp.org/cms/index.cfm?fa=view&id=3252.

33. Data Source: FRED, Federal Reserve Economic Data, Federal Reserve Bank of St. Louis: All Employees: Total Nonfarm, Thousands of Persons, Seasonally Adjusted [PAYEMS]; US Department of Labor: Bureau of Labor Statistics; https://research.stlouisfed.org/fred2/series/PAYEMS/; accessed June 10, 2014.

34. Ghayad's experiment is described in Mathew O'Brien, "The Terrifying Reality of Long-Term Unemployment," *The Atlantic,* April 13, 2013, http://www.theatlantic.com/business/archive/2013/04/the-terrifying-reality-of-long-term-unemployment/274957/.

35. Regarding the Urban Institute report on long-term unemployment, see Mathew O'Brien, "Who Are the Long-Term Unemployed?," *The Atlantic,* August 23, 2013, http://www.theatlantic.com/business/archive/2013/08/who-are-the-long-term-unemployed/278964, and Josh Mitchell, "Who Are the Long-Term Unemployed?," Urban Institute, July 2013, http://www.urban.org/uploadedpdf/412885-who-are-the-long-term-unemployed.pdf.

36. "The Gap Widens Again," *The Economist,* March 10, 2012, http://www.economist.com/node/21549944.

37. Emmanuel Saez, "Striking It Richer: The Evolution of Top Incomes in the United States," University of California, Berkeley, September 3, 2013, http://elsa.berkeley.edu/~saez/saez-UStopincomes-2012.pdf.

38. CIA World Factbook, "Country Comparison: Distribution of Family Income: Gini Index," https://www.cia.gov/library/publications/the-world-factbook/rankorder/2172rank.html; accessed April 29, 2014.

39. Dan Ariely, "Americans Want to Live in a Much More Equal Country (They Just Don't Realize It)," *The Atlantic,* August 2, 2012, http://www.theatlantic.com/business/archive/2012/08/americans-want-to-live-in-a-much-more-equal-country-they-just-dont-realize-it/260639/.

40. Jonathan James, "The College Wage Premium," Federal Reserve Bank of Cleveland, Economic Commentary, August 8, 2012, http://www.clevelandfed.org/research/commentary/2012/2012–10.cfm.

41. Diana G. Carew, "No Recovery for Young People" (Progressive Policy Institute blog), August 5, 2013, http://www.progressivepolicy.org/2013/08/no-recovery-for-young-people/.

42. Nir Jaimovich and Henry E. Siu, "The Trend Is the Cycle: Job Polarization and Jobless Recoveries," National Bureau of Economic Research, Working Paper No. 18334, issued in August 2012, http://www.nber.org/papers/w18334, also available at http://faculty.arts.ubc.ca/hsiu/research/polar20120331.pdf.

43. See, for example, Ben Casselman, "Low Pay Clouds Job Growth," *Wall Street Journal,* April 3, 2013, http://online.wsj.com/article/SB10001424127887324635904578643654030630378.html.

44. This information comes from the Bureau of Labor Statistics monthly employment reports. The December 2007 report (http://www.bls.gov/news.release/archives/empsit_01042008.pdf), Table A-5, shows 122 million full-time jobs and about 24 million part-time jobs. The August 2013 report (http://www.bls.gov/news.release/archives/empsit_09062013.pdf), Table A-8, shows about 117 million full-time jobs and 27 million part-time jobs.

45. David Autor, "The Polarization of Job Opportunities in the U.S. Labor Market: Implications for Employment and Earnings," a paper jointly released by The Center for American Progress and The Hamilton Project, April 2010, pp. 8–9, http://economics.mit.edu/files/5554.

46. Ibid., p. 4.

47. Ibid., p. 2.

48. Jaimovich and Siu, "The Trend Is the Cycle: Job Polarization and Jobless Recoveries," p. 2.

49. Chrystia Freeland, "The Rise of 'Lovely' and 'Lousy' Jobs," Reuters, April 12, 2012, http://www.reuters.com/article/2012/04/12/column-freeland-middleclass-idUSL2E8FCCZZ20120412.

50. Galina Hale and Bart Hobijn, "The U.S. Content of 'Made in China,'" Federal Reserve Bank of San Francisco (FRBSF), Economic Letter, August 8, 2011, http://www.frbsf.org/economic-research/publications/economic-letter/2011/august/us-made-in-china/.

51. Data Source: FRED, Federal Reserve Economic Data, Federal Reserve Bank of St. Louis: All Employees Manufacturing, Thousands of Persons, Seasonally Adjusted [MANEMP] divided by All Employees: Total Nonfarm, Thousands of Persons, Seasonally Adjusted [PAYEMS]; US Department of Labor: Bureau of Labor Statistics; https://research.stlouisfed.org/fred2/series/PAYEMS/; accessed June 10, 2014.

52. Bruce Bartlett, "'Financialization' as a Cause of Economic Malaise," *New York Times* (Economix blog), June 11, 2013, http://economix.blogs.nytimes.com/2013/06/11/financialization-as-a-cause-of-economic-malaise/; Brad Delong, "The Financialization of the American Economy" (blog), October 18, 2011, http://delong.typepad.com/sdj/2011/10/the-financialization-of-the-american-economy.html.

53. Simon Johnson and James Kwak, *13 Bankers: The Wall Street Takeover and the Next Financial Meltdown* (New York: Pantheon, 2010), pp. 85–86.

54. Matt Taibbi, "The Great American Bubble Machine," *Rolling Stone*, July 9, 2009, http://www.rollingstone.com/politics/news/the-great-american-bubble-machine-20100405.

55. There are a number of economic papers demonstrating the relationship between financialization and inequality. For a comprehensive treatment, see James K. Galbraith, *Inequality and Instability: A Study of the World Economy Just Before the Great Crisis* (New York: Oxford University Press, 2012). On the relationship between financialization and the decline in labor's share, see *Global Wage Report 2012/13*, International Labour Organization, 2013, http://www.ilo.org/wcmsp5/groups/public/—dgreports/—dcomm/—publ/documents/publication/wcms_194843.pdf.

56. Susie Poppick, "4 Ways the Market Could *Really* Surprise You," *CNN Money*, January 28, 2013, http://money.cnn.com/gallery/investing/2013/01/28/stock-market-crash.moneymag/index.html.

57. Matthew Yglesias, "America's Private Sector Labor Unions Have Always Been in Decline," *Slate* (Moneybox blog), March 20, 2013, http://www.slate.com/blogs/moneybox/2013/03/20/private_sector_labor_unions_have_always_been_in_decline.html.

58. On Canadian median wages and unionization, see Miles Corak, "The Simple Economics of the Declining Middle Class—and the Not So Simple Politics," *Economics for Public Policy Blog*, August 7, 2013, http://milescorak.com/2013/08/07/the-simple-economics-of-the-declining-middle-class-and-the-not-so-simple-politics/, and "Unions on Decline in Private Sector," *CBC News Canada*, September 2, 2012, http://www.cbc.ca/news/canada/unions-on-decline-in-private-sector-1.1150562.

59. Carl Benedikt Frey and Michael A. Osborne, "The Future of Employment: How Susceptible Are Jobs to Computerisation?," Oxford Martin School, Programme on the Impacts of Future Technology, September 17, 2013, p. 38, http://www.futuretech.ox.ac.uk/sites/futuretech.ox.ac.uk/files/The_Future_of_Employment_OMS_Working_Paper_1.pdf.

60. Paul Krugman, "Robots and Robber Barons," *New York Times,* December 9, 2012, http://www.nytimes.com/2012/12/10/opinion/krugman-robots-and-robber-barons.html?gwh=054BD73AB17F28CD31B3999AABFD7E86; Jeffrey D. Sachs and Laurence J. Kotlikoff, "Smart Machines and Long-Term Misery," National Bureau of Economic Research, Working Paper No. 18629, issued in December 2012, http://www.nber.org/papers/w18629.pdf.

Chapter 3

1. Robert J. Gordon, "Is U.S. Economic Growth Over? Faltering Innovation Confronts the Six Headwinds," National Bureau of Economic Research, NBER Working Paper 18315, issued in August 2012, http://www.nber.org/papers/w18315; see also http://faculty-web.at.northwestern.edu/economics/gordon/is%20us%20economic%20growth%20over.pdf.

2. For a more detailed explanation of semiconductor fabrication S-curves, see Murrae J. Bowden, "Moore's Law and the Technology S-Curve," *Current Issues in Technology Management,* Stevens Institute of Technology, Winter 2004, https://www.stevens.edu/howe/sites/default/files/bowden_0.pdf.

3. See, for example, Michael Kanellos, "With 3D Chips, Samsung Leaves Moore's Law Behind," *Forbes.com,* August 14, 2013, http://www.forbes.com/sites/michaelkanellos/2013/08/14/with-3d-chips-samsung-leaves-moores-law-behind; John Markoff, "Researchers Build a Working Carbon Nanotube Computer," *New York Times,* September 25, 2013, http://www.nytimes.com/2013/09/26/science/researchers-build-a-working-carbon-nanotube-computer.html?ref=johnmarkoff&_r=0.

4. President's Council of Advisors on Science and Technology, "Report to the President and Congress: Designing a Digital Future: Federally Funded Research and Development in Networking and Information Technology," December 2010, p. 71, http://www.whitehouse.gov/sites/default/files/microsites/ostp/pcast-nitrd-report-2010.pdf.

5. James Fallows, "Why Is Software So Slow?," *The Atlantic,* August 14, 2013, http://www.theatlantic.com/magazine/archive/2013/09/why-is-software-so-slow/309422/.

6. Science writer Joy Casad calculates that the speed at which neurons transmit signals is about half a millisecond. That is vastly slower than what occurs in computer chips. See Joy Casad, "How Fast Is a Thought?," *Examiner.com,* August 20, 2009, http://www.examiner.com/article/how-fast-is-a-thought.

7. IBM Press Release: "IBM Research Creates New Foundation to Program SyNAPSE Chips," August 8, 2013, http://finance.yahoo.com/news/ibm-research-creates-foundation-program-040100103.html.

8. See, for example, "Rise of the Machines," *The Economist* (Free Exchange blog), October 20, 2010, http://www.economist.com/blogs/freeexchange/2010/10/technology.

9. Google Investor Relations website, http://investor.google.com/financial/tables.html.

10. Historical data on General Motors can be found at http://money.cnn.com/magazines/fortune/fortune500_archive/snapshots/1979/563.html. GM earned $3.5 billion in 1979, which is equivalent to about $11 billion in 2012 dollars.

11. Scott Timberg, "Jaron Lanier: The Internet Destroyed the Middle Class," *Salon.com,* May 12, 2013, http://www.salon.com/2013/05/12/jaron_lanier_the_internet_destroyed_the_middlc_class/.

12. This video can be found at https://www.youtube.com/watch?v=wb2cI_gJUok, or search YouTube for "Man vs. Machine: Will Human Workers Become Obsolete?" Kurzweil's remarks can be found at about 05:40.

13. Robert Jensen, "The Digital Provide: Information (Technology), Market Performance and Welfare in the South Indian Fisheries Sector," *Quarterly Journal of Economics,* 122, no. 3 (2007): 879–924.

14. A few places in which the story of the sardine fishermen of Kerala has been told are *The Rational Optimist* by Matt Ridley, *A History of the World in 100 Objects* by Neil MacGregor, *The Mobile Wave* by Michael Saylor, *Race Against the Machine* by Erik Brynjolfsson and Andrew McAfee, *Content Nation* by John Blossom, *Planet India* by Mira Kamdar, and "To Do with the Price of Fish," *The Economist,* May 10, 2007. And now this book joins the list.

Chapter 4

1. David Carr, "The Robots Are Coming! Oh, They're Here," *New York Times* (Media Decoder blog), October 19, 2009, http://mediadecoder.blogs.nytimes.com/2009/10/19/the-robots-are-coming-oh-theyre-here.

2. Steven Levy, "Can an Algorithm Write a Better News Story Than a Human Reporter?," *Wired,* April 24, 2012, http://www.wired.com/2012/04/can-an-algorithm-write-a-better-news-story-than-a-human-reporter.

3. Narrative Science corporate website, http://narrativescience.com.

4. George Leef, "The Skills College Graduates Need," Pope Center for Education Policy, December 14, 2006, http://www.popecenter.org/commentaries/article.html?id=1770.

5. Kenneth Neil Cukier and Viktor Mayer-Schoenberger, "The Rise of Big Data," *Foreign Affairs,* May/June 2013, http://www.foreignaffairs.com

/articles/139104/kenneth-neil-cukier-and-viktor-mayer-schoenberger/the-rise-of-big-data.

6. Thomas H. Davenport, Paul Barth, and Randy Bean, "How 'Big Data' Is Different," *MIT Sloan Management Review,* July 20, 2012, http://sloanreview.mit.edu/article/how-big-data-is-different.

7. Charles Duhigg, "How Companies Learn Your Secrets," *New York Times,* February 16, 2012, http://www.nytimes.com/2012/02/19/magazine/shopping-habits.html.

8. As quoted in Steven Levy, *In the Plex: How Google Thinks, Works, and Shapes Our Lives* (New York: Simon and Schuster, 2011), p. 64.

9. Tom Simonite, "Facebook Creates Software That Matches Faces Almost as Well as You Do," *MIT Technology Review,* March 17, 2014, http://www.technologyreview.com/news/525586/facebook-creates-software-that-matches-faces-almost-as-well-as-you-do/.

10. As quoted in John Markoff, "Scientists See Promise in Deep-Learning Programs," *New York Times,* November 23, 2012, http://www.nytimes.com/2012/11/24/science/scientists-see-advances-in-deep-learning-a-part-of-artificial-intelligence.html.

11. Don Peck, "They're Watching You at Work," *The Atlantic,* December 2013, http://www.theatlantic.com/magazine/archive/2013/12/theyre-watching-you-at-work/354681/.

12. United States Patent No. 8,589,407, "Automated Generation of Suggestions for Personalized Reactions in a Social Network," November 19, 2013, http://patft.uspto.gov/netacgi/nph-Parser?Sect1=PTO2&Sect2=HITOFF&p=1&u=%2Fnetahtml%2FPTO%2Fsearch-adv.htm&r=1&f=G&l=50&d=PALL&S1=08589407&OS=PN/08589407&RS=PN/08589407.

13. This WorkFusion information is based on a telephone conversation between the author and Adam Devine, vice president of Product Marketing & Strategic Partnerships at WorkFusion, on May 14, 2014.

14. This incident is recounted in Steven Baker, *Final Jeopardy: Man vs. Machine and the Quest to Know Everything* (New York: Houghton Mifflin Harcourt, 2011), p. 20. The story of the steakhouse dinner is also told in John E. Kelly III, *Smart Machines: IBM's Watson and the Era of Cognitive Computing* (New York: Columbia University Press, 2013), p. 27. However, Baker's book indicates that some IBM employees believe the idea to build a *Jeopardy!*-playing computer predates the dinner.

15. Rob High, "The Era of Cognitive Systems: An Inside Look at IBM Watson and How it Works," *IBM Redbooks,* 2012, p. 2, http://www.redbooks.ibm.com/redpapers/pdfs/redp4955.pdf.

16. Baker, *Final Jeopardy: Man vs. Machine and the Quest to Know Everything,* p. 30.

17. Ibid., pp. 9 and 26.

18. Ibid., p. 68.
19. Ibid.
20. Ibid., p. 78.
21. David Ferrucci, Eric Brown, Jennifer Chu-Carroll, James Fan, David Gondek, Aditya A. Kalyanpur, Adam Lally, J. William Murdock, Eric Nyberg, John Prager, Nico Schlaefer, and Chris Welty, "Building Watson: An Overview of the DeepQA Project," *AI Magazine,* Fall 2010, http://www.aaai.org/Magazine/Watson/watson.php.
22. IBM Press Release: "IBM Research Unveils Two New Watson Related Projects from Cleveland Clinic Collaboration," October 15, 2013, http://www-03.ibm.com/press/us/en/pressrelease/42203.wss.
23. IBM Case Study: "IBM Watson/Fluid, Inc.," November 4, 2013, http://www-03.ibm.com/innovation/us/watson/pdf/Fluid_case_study_11_4_2013.pdf.
24. "IBM Watson/MD Buyline, Inc.," IBM Case Study, November 4, 2013, http://www-03.ibm.com/innovation/us/watson/pdf/MDB_case_study_11_4_2013.pdf.
25. IBM Press Release: "Citi and IBM Enter Exploratory Agreement on Use of Watson Technologies," March 5, 2012, http://www-03.ibm.com/press/us/en/pressrelease/37029.wss.
26. IBM Press Release: "IBM Watson's Next Venture: Fueling New Era of Cognitive Apps Built in the Cloud by Developers," November 14, 2013, http://www-03.ibm.com/press/us/en/pressrelease/42451.wss.
27. Quentin Hardy, "IBM to Announce More Powerful Watson via the Internet," *New York Times,* November 13, 2013, http://www.nytimes.com/2013/11/14/technology/ibm-to-announce-more-powerful-watson-via-the-internet.html?_r=0.
28. Nick Heath, "'Let's Try and Not Have a Human Do It': How One Facebook Techie Can Run 20,000 Servers," *ZDNet,* November 25, 2013, http://www.zdnet.com/lets-try-and-not-have-a-human-do-it-how-one-facebook-techie-can-run-20000-servers-7000023524.
29. Michael S. Rosenwald, "Cloud Centers Bring High-Tech Flash But Not Many Jobs to Beaten-Down Towns," *Washington Post,* November 24, 2011, http://www.washingtonpost.com/business/economy/cloud-centers-bring-high-tech-flash-but-not-many-jobs-to-beaten-down-towns/2011/11/08/gIQAccTQtN_print.html.
30. Quentin Hardy, "Active in Cloud, Amazon Reshapes Computing," *New York Times,* August 27, 2012, http://www.nytimes.com/2012/08/28/technology/active-in-cloud-amazon-reshapes-computing.html.
31. Mark Stevenson, *An Optimist's Tour of the Future: One Curious Man Sets Out to Answer "What's Next?"* (New York: Penguin Group, 2011), p. 101.
32. Michael Schmidt and Hod Lipson, "Distilling Free-Form Natural Laws from Experimental Data," *Science* 324 (April 3, 2009), http://creativemachines.cornell.edu/sites/default/files/Science09_Schmidt.pdf.

33. Stevenson, *An Optimist's Tour of the Future*, p. 104.

34. National Science Foundation Press Release: "Maybe Robots Dream of Electric Sheep, But Can They Do Science?," April 2, 2009, http://www.nsf.gov/mobile/news/news_summ.jsp?cntn_id=114495.

35. Asaf Shtull-Trauring, "An Israeli Professor's 'Eureqa' Moment," *Haaretz*, February 3, 2012, http://www.haaretz.com/weekend/magazine/an-israeli-professor-s-eureqa-moment-1.410881.

36. John R. Koza, "Human-Competitive Results Produced by Genetic Programming," *Genetic Programming and Evolvable Machines* 11, nos. 3–4 (September 2010), http://dl.acm.org/citation.cfm?id=1831232.

37. John Koza's website, http://www.genetic-programming.com/#_What_is_Genetic, also: http://eventful.com/events/john-r-koza-routine-human-competitive-machine-intelligence-/E0–001–000292572–0.

38. Lev Grossman, "2045: The Year Man Becomes Immortal," *Time*, February 10, 2011, http://content.time.com/time/magazine/article/0,9171,2048299,00.html.

39. As quoted in Sylvia Smith, "Iamus: Is This the 21st Century's Answer to Mozart?," *BBC News*, January 2, 2013, http://www.bbc.co.uk/news/technology-20889644.

40. As quoted in Kadim Shubber, "Artificial Artists: When Computers Become Creative," *Wired Magazine–UK*, August 13, 2007, http://www.wired.co.uk/news/archive/2013–08/07/can-computers-be-creative/viewgallery/306906.

41. Shubber, "Artificial Artists: When Computers Become Creative."

42. "Bloomberg Bolsters Machine-Readable News Offering," *The Trade*, February 19, 2010, http://www.thetradenews.com/News/Operations_Technology/Market_data/Bloomberg_bolsters_machine-readable_news_offering.aspx.

43. Neil Johnson, Guannan Zhao, Eric Hunsader, Hong Qi, Nicholas Johnson, Jing Meng, and Brian Tivnan, "Abrupt Rise of New Machine Ecology Beyond Human Response Time," *Nature*, September 11, 2013, http://www.nature.com/srep/2013/130911/srep02627/full/srep02627.html.

44. Christopher Steiner, *Automate This: How Algorithms Came to Rule Our World* (New York: Portfolio/Penguin, 2012), pp. 116–120.

45. Max Raskin and Ilan Kolet, "Wall Street Jobs Plunge as Profits Soar: Chart of the Day," *Bloomberg News*, April 23, 2013, http://www.bloomberg.com/news/2013–04–24/wall-street-jobs-plunge-as-profits-soar-chart-of-the-day.html.

46. Steve Lohr, "David Ferrucci: Life After Watson," *New York Times* (Bits blog), May 6, 2013, http://bits.blogs.nytimes.com/2013/05/06/david-ferrucci-life-after-watson/?_r=1.

47. As quoted in Alan S. Blinder, "Offshoring: The Next Industrial Revolution?," *Foreign Affairs*, March/April 2006, http://www.foreignaffairs.com/articles/61514/alan-s-blinder/offshoring-the-next-industrial-revolution.

48. Alan S. Blinder, "Free Trade's Great, but Offshoring Rattles Me," *Washington Post,* May 6, 2007, http://www.washingtonpost.com/wp-dyn/content/article/2007/05/04/AR2007050402555.html.

49. Blinder, "Offshoring: The Next Industrial Revolution?"

50. Carl Benedikt Frey and Michael A. Osborne, "The Future of Employment: How Susceptible Are Jobs to Computerisation?," Oxford Martin School, Programme on the Impacts of Future Technology, September 17, 2013, p. 38, http://www.futuretech.ox.ac.uk/sites/futuretech.ox.ac.uk/files/The_Future_of_Employment_OMS_Working_Paper_1.pdf.

51. Alan S. Blinder, "On the Measurability of Offshorability," *VOX,* October 9, 2009, http://www.voxeu.org/article/twenty-five-percent-us-jobs-are-offshorable.

52. Keith Bradsher, "Chinese Graduates Say No Thanks to Factory Jobs," *New York Times,* January 24, 2013, http://www.nytimes.com/2013/01/25/business/as-graduates-rise-in-china-office-jobs-fail-to-keep-up.html; Keith Bradsher, "Faltering Economy in China Dims Job Prospects for Graduates," *New York Times,* June 16, 2013, http://www.nytimes.com/2013/06/17/business/global/faltering-economy-in-china-dims-job-prospects-for-graduates.html?pagewanted=all.

53. Eric Mack, "Google Has a 'Near Perfect' Universal Translator—for Portuguese, at Least," *CNET News,* July 28, 2013, http://news.cnet.com/8301–17938_105–57595825–1/google-has-a-near-perfect-universal-translator-for-portuguese-at-least/.

54. Tyler Cowen, *Average Is Over: Powering America Beyond the Age of the Great Stagnation* (New York: Dutton, 2013), p. 79.

55. John Markoff, "Armies of Expensive Lawyers, Replaced by Cheaper Software," *New York Times,* March 4, 2011, http://www.nytimes.com/2011/03/05/science/05legal.html.

56. Arin Greenwood, "Attorney at Blah," *Washington City Paper,* November 8, 2007, http://www.washingtoncitypaper.com/articles/34054/attorney-at-blah.

57. Erin Geiger Smith, "Shocking? Temp Attorneys Must Review 80 Documents Per Hour," *Business Insider,* October 21, 2009, http://www.businessinsider.com/temp-attorney-told-to-review-80-documents-per-hour-2009–10.

58. Ian Ayres, *Super Crunchers: Why Thinking in Numbers Is the New Way to Be Smart* (New York: Bantam Books, 2007), p. 117.

59. "Peter Thiel's Graph of the Year," *Washington Post* (Wonkblog), December 30, 2013, http://www.washingtonpost.com/blogs/wonkblog/wp/2013/12/30/peter-thiels-graph-of-the-year/.

60. Paul Beaudry, David A. Green, and Benjamin M. Sand, "The Great Reversal in the Demand for Skill and Cognitive Tasks," National Bureau of

Economic Research, NBER Working Paper No. 18901, issued in March 2013, http://www.nber.org/papers/w18901.

61. Hal Salzman, Daniel Kuehn, and B. Lindsay Lowell, "Guestworkers in the High-Skill U.S. Labor Market," Economic Policy Institute, April 24, 2013, http://www.epi.org/publication/bp359-guestworkers-high-skill-labor-market-analysis/.

62. As quoted in Michael Fitzpatrick, "Computers Jump to the Head of the Class," *New York Times,* December 29, 2013, http://www.nytimes.com/2013/12/30/world/asia/computers-jump-to-the-head-of-the-class.html.

Chapter 5

1. This petition can be viewed at http://humanreaders.org/petition/.

2. University of Akron News Release: "Man and Machine: Better Writers, Better Grades," April 12, 2012, http://www.uakron.edu/education/about-the-college/news-details.dot?newsId=40920394–9e62–415d-b038–15fe2e72a677&pageTitle=Recent%20Headlines&crumbTitle=Man%20and%20%20machine:%20Better%20writers,%20better%20grades.

3. Ry Rivard, "Humans Fight over Robo-Readers," *Inside Higher Ed,* March 15, 2013, http://www.insidehighered.com/news/2013/03/15/professors-odds-machine-graded-essays.

4. John Markoff, "Essay-Grading Software Offers Professors a Break," *New York Times,* April 4, 2013, http://www.nytimes.com/2013/04/05/science/new-test-for-computers-grading-essays-at-college-level.html.

5. John Markoff, "Virtual and Artificial, but 58,000 Want Course," *New York Times,* August 15, 2011, http://www.nytimes.com/2011/08/16/science/16stanford.html?_r=0.

6. The story of the Stanford AI course is drawn from Max Chafkin, "Udacity's Sebastian Thrun, Godfather of Free Online Education, Changes Course," *Fast Company,* December 2013/January 2014, http://www.fastcompany.com/3021473/udacity-sebastian-thrun-uphill-climb; Jeffrey J. Selingo, *College Unbound: The Future of Higher Education and What It Means for Students* (New York: New Harvest, 2013), pp. 86–101; and Felix Salmon, "Udacity and the Future of Online Universities" (Reuters blog), January 23, 2012, http://blogs.reuters.com/felix-salmon/2012/01/23/udacity-and-the-future-of-online-universities/.

7. Thomas L. Friedman, "Revolution Hits the Universities," *New York Times,* January 26, 2013, http://www.nytimes.com/2013/01/27/opinion/sunday/friedman-revolution-hits-the-universities.html.

8. Penn Graduate School of Education Press Release: "Penn GSE Study Shows MOOCs Have Relatively Few Active Users, with Only a Few Persisting to Course End," December 5, 2013, http://www.gse.upenn.edu/pressroom/press-releases/2013/12/penn-gse-study-shows-moocs-have-relatively-few-active-users-only-few-persisti.

9. Tamar Lewin, "After Setbacks, Online Courses Are Rethought," *New York Times,* December 10, 2013, http://www.nytimes.com/2013/12/11/us/after-setbacks-online-courses-are-rethought.html.

10. Alexandra Tilsley, "Paying for an A," *Inside Higher Ed,* September 21, 2012, http://www.insidehighered.com/news/2012/09/21/sites-offering-take-courses-fee-pose-risk-online-ed.

11. Jeffrey R. Young, "Dozens of Plagiarism Incidents Are Reported in Coursera's Free Online Courses," *Chronicle of Higher Education,* August 16, 2012, http://chronicle.com/article/Dozens-of-Plagiarism-Incidents/133697/.

12. "MOOCs and Security" (MIT Geospacial Data Center blog), October 9, 2012, http://cybersecurity.mit.edu/2012/10/moocs-and-security/.

13. Steve Kolowich, "Doubts About MOOCs Continue to Rise, Survey Finds," *Chronicle of Higher Education,* January 15, 2014, http://chronicle.com/article/Doubts-About-MOOCs-Continue-to/144007/.

14. Jeffrey J. Selingo, *College Unbound,* p. 4.

15. Michelle Jamrisko and Ilan Kole, "College Costs Surge 500% in U.S. Since 1985: Chart of the Day," *Bloomberg Personal Finance,* August 26, 2013, http://www.bloomberg.com/news/2013–08–26/college-costs-surge-500-in-u-s-since-1985-chart-of-the-day.html.

16. On student loans, see Rohit Chopra, "Student Debt Swells, Federal Loans Now Top a Trillion," Consumer Financial Protection Bureau, July 17, 2013, and Blake Ellis, "Average Student Loan Debt: $29,400," *CNN Money,* December 5, 2013, http://money.cnn.com/2013/12/04/pf/college/student-loan-debt/.

17. This information regarding graduation rates is provided by the National Center of Education Statistics, http://nces.ed.gov/fastfacts/display.asp?id=40.

18. Selingo, *College Unbound,* p. 27.

19. "Senior Administrators Now Officially Outnumber Faculty at the UC" (Reclaim UC blog), September 19, 2011, http://reclaimuc.blogspot.com/2011/09/senior-administrators-now-officially.html.

20. Selingo, *College Unbound,* p. 28.

21. Ibid.

22. Clayton Christensen interview with Mark Suster at Startup Grind 2013, available at YouTube, http://www.youtube.com/watch?v=KYVdf5xyD8I.

23. William G. Bowen, Matthew M. Chingos, Kelly A. Lack, and Thomas I. Nygren, "Interactive Learning Online at Public Universities: Evidence from Randomized Trials," *Ithaka S+R Research Publication,* May 22, 2012, http://www.sr.ithaka.org/research-publications/interactive-learning-online-public-universities-evidence-randomized-trials.

Chapter 6

1. These two cases of cobalt poisoning were reported by Gina Kolata, "As Seen on TV, a Medical Mystery Involving Hip Implants Is Solved," *New York Times,* February 6, 2014, http://www.nytimes.com/2014/02/07/health/house-plays-a-role-in-solving-a-medical-mystery.html.

2. Catherine Rampell, "U.S. Health Spending Breaks from the Pack," *New York Times* (Economix blog), July 8, 2009, http://economix.blogs.nytimes.com/2009/07/08/us-health-spending-breaks-from-the-pack/.

3. IBM corporate website, http://www-03.ibm.com/innovation/us/watson/watson_in_healthcare.shtml.

4. Spencer E. Ante, "IBM Struggles to Turn Watson Computer into Big Business," *Wall Street Journal,* January 7, 2014, http://online.wsj.com/news/articles/SB10001424052702304887104579306881917668654.

5. Dr. Courtney DiNardo, as quoted in Laura Nathan-Garner, "The Future of Cancer Treatment and Research: What IBM Watson Means for Our Patients," *MD Anderson—Cancerwise,* November 12, 2013, http://www2.mdanderson.org/cancerwise/2013/11/the-future-of-cancer-treatment-and-research-what-ibm-watson-means-for-patients.html.

6. Mayo Clinic Press Release: "Artificial Intelligence Helps Diagnose Cardiac Infections," September 12, 2009, http://www.eurekalert.org/pub_releases/2009–09/mc-aih090909.php.

7. National Research Council, *Preventing Medication Errors: Quality Chasm Series* (Washington, DC: National Academies Press, 2007), p. 47.

8. National Research Council, *To Err Is Human: Building a Safer Health System* (Washington, DC: National Academies Press, 2000), p. 1.

9. National Academies News Release: "Medication Errors Injure 1.5 Million People and Cost Billions of Dollars Annually," July 20, 2006, http://www8.nationalacademies.org/onpinews/newsitem.aspx?RecordID=11623.

10. Martin Ford, "Dr. Watson: How IBM's Supercomputer Could Improve Health Care," *Washington Post,* September 16, 2011, http://www.washingtonpost.com/opinions/dr-watson-how-ibms-supercomputer-could-improve-health-care/2011/09/14/gIQAOZQzXK_story.html.

11. Roger Stark, "The Looming Doctor Shortage," Washington Policy Center, November 2011, http://www.washingtonpolicy.org/publications/notes/looming-doctor-shortage.

12. Marijke Vroomen Durning, "Automated Breast Ultrasound Far Faster Than Hand-Held," *Diagnostic Imaging,* May 3, 2012, http://www.diagnosticimaging.com/articles/automated-breast-ultrasound-far-faster-hand-held.

13. On the "double reading" strategy in radiology, see Farhad Manjoo, "Why the Highest-Paid Doctors Are the Most Vulnerable to Automation," *Slate,*

September 27, 2011, http://www.slate.com/articles/technology/robot_invasion/2011/09/will_robots_steal_your_job_3.html; I. Anttinen, M. Pamilo, M. Soiva, and M. Roiha, "Double Reading of Mammography Screening Films—One Radiologist or Two?," *Clinical Radiology* 48, no. 6 (December 1993): 414–421, http://www.ncbi.nlm.nih.gov/pubmed/8293648?report=abstract; and Fiona J. Gilbert et al., "Single Reading with Computer-Aided Detection for Screening Mammography," *New England Journal of Medicine,* October 16, 2008, http://www.nejm.org/doi/pdf/10.1056/NEJMoa0803545.

14. Manjoo, "Why the Highest-Paid Doctors Are the Most Vulnerable to Automation."

15. Rachael King, "Soon, That Nearby Worker Might Be a Robot," *Bloomberg Businessweek,* June 2, 2010, http://www.businessweek.com/stories/2010–06–02/soon-that-nearby-worker-might-be-a-robotbusinessweek-business-news-stock-market-and-financial-advice.

16. GE Corporate Press Release: "GE to Develop Robotic-Enabled Intelligent System Which Could Save Patients Lives and Hospitals Millions," January 30, 2013, http://www.genewscenter.com/Press-Releases/GE-to-Develop-Robotic-enabled-Intelligent-System-Which-Could-Save-Patients-Lives-and-Hospitals-Millions-3dc2.aspx.

17. I-Sur website, http://www.isur.eu/isur/.

18. Statistics on US aging can be found at the Department of Health and Human Services' Administration on Aging website, http://www.aoa.gov/Aging_Statistics/.

19. For statistics on Japanese aging, see "Difference Engine: The Caring Ro-bot," *The Economist,* May 14, 2014, http://www.economist.com/blogs/babbage/2013/05/automation-elderly.

20. Ibid.

21. "Robotic Exoskeleton Gets Safety Green Light," *Discovery News,* February 27, 2013, http://news.discovery.com/tech/robotics/robotic-exoskeleton-gets-safety-green-light-130227.htm.

22. US Bureau of Labor Statistics, *Occupational Outlook Handbook,* http://www.bls.gov/ooh/most-new-jobs.htm.

23. Heidi Shierholz, "Six Years from Its Beginning, the Great Recession's Shadow Looms over the Labor Market," Economic Policy Institute, January 9, 2014, http://www.epi.org/publication/years-beginning-great-recessions-shadow/.

24. Steven Brill, "Bitter Pill: How Outrageous and Egregious Profits Are Destroying Our Health Care," *Time,* March 4, 2013.

25. Elisabeth Rosenthal, "As Hospital Prices Soar, a Stitch Tops $500," *New York Times,* December 2, 2013, http://www.nytimes.com/2013/12/03/health/as-hospital-costs-soar-single-stitch-tops-500.html.

26. Kenneth J. Arrow, "Uncertainty and the Welfare Economics of Medical Care," *American Economic Review,* December 1963, http://www.who.int/bulletin/volumes/82/2/PHCBP.pdf.

27. "The Concentration of Health Care Spending: NIHCM Foundation Data Brief July 2012," National Institute for Health Care Management, July 2012, http://nihcm.org/images/stories/DataBrief3_Final.pdf.

28. Brill, "Bitter Pill."

29. Jenny Gold, "Proton Beam Therapy Heats Up Hospital Arms Race," *Kaiser Health News,* May 2013, http://www.kaiserhealthnews.org/stories/2013/may/31/proton-beam-therapy-washington-dc-health-costs.aspx.

30. James B. Yu, Pamela R. Soulos, Jeph Herrin, Laura D. Cramer, Arnold L. Potosky, Kenneth B. Roberts, and Cary P. Gross, "Proton Versus Intensity-Modulated Radiotherapy for Prostate Cancer: Patterns of Care and Early Toxicity," *Journal of the National Cancer Institute* 105, no. 1 (January 2, 2013), http://jnci.oxfordjournals.org/content/105/1.toc.

31. Gold, "Proton Beam Therapy Heats Up Hospital Arms Race."

32. Sarah Kliff, "Maryland's Plan to Upend Health Care Spending," *Washington Post* (Wonkblog), January 10, 2014, http://www.washingtonpost.com/blogs/wonkblog/wp/2014/01/10/%253Fp%253D74854/.

33. "Underpayment by Medicare and Medicaid Fact Sheet," American Hospital Association, December 2010, http://www.aha.org/content/00–10/10medunderpayment.pdf.

34. Ed Silverman, "Increased Abandonment of Prescriptions Means Less Control of Chronic Conditions," *Managed Care,* June 2010, http://www.managedcaremag.com/archives/1006/1006.abandon.html.

35. Dean Baker, "Financing Drug Research: What Are the Issues?," Center for Economic and Policy Research, September 2004, http://www.cepr.net/index.php/Publications/Reports/financing-drug-research-what-are-the-issues.

36. Matthew Perrone, "Scooter Ads Face Scrutiny from Gov't, Doctors," Associated Press, March 28, 2013, http://news.yahoo.com/scooter-ads-face-scrutiny-govt-doctors-141816931–finance.html.

37. Farhad Manjoo, "My Father the Pharmacist vs. a Gigantic Pill-Packing Machine," *Slate,* http://www.slate.com/articles/technology/robot_invasion/2011/09/will_robots_steal_your_job_2.html.

38. Daniel L. Brown, "A Looming Joblessness Crisis for New Pharmacy Graduates and the Implications It Holds for the Academy," *American Journal of Pharmacy Education* 77, no. 5 (June 13, 2012): 90, http://www.ncbi.nlm.nih.gov/pmc/articles/PMC3687123/.

Chapter 7

1. GE's corporate website, https://www.ge.com/stories/additive-manufacturing.

2. American Airlines News Release: "American Becomes the First Major Commercial Carrier to Deploy Electronic Flight Bags Throughout Fleet and Discontinue Paper Revisions," June 24, 2013, http://hub.aa.com/en/nr/pressrelease/american-airlines-completes-electronic-flight-bag-implementation.

3. Tim Catts, "GE Turns to 3D Printers for Plane Parts," *Bloomberg Businessweek*, November 27, 2013, http://www.businessweek.com/articles/2013–11–27/general-electric-turns-to-3d-printers-for-plane-parts.

4. Lucas Mearian, "The First 3D Printed Organ—a Liver—Is Expected in 2014," *ComputerWorld*, December 26, 2013, http://www.computerworld.com/s/article/9244884/The_first_3D_printed_organ_a_liver_is_expected_in_2014?taxonomyId=128&pageNumber=2.

5. Hod Lipson and Melba Kurman, *Fabricated: The New World of 3D Printing* (New York, John Wiley & Sons, 2013).

6. Mark Hattersley, "The 3D Printer That Can Build a House in 24 Hours," *MSN Innovation*, November 11, 2013, http://innovation.uk.msn.com/design/the-3d-printer-that-can-build-a-house-in-24-hours.

7. Information regarding construction employment in the United States can be found at the US Bureau of Labor Statistics website: http://www.bls.gov/iag/tgs/iag23.htm.

8. Further details are available at the DARPA Grand Challenge website: http://archive.darpa.mil/grandchallenge/.

9. Tom Simonite, "Data Shows Google's Robot Cars Are Smoother, Safer Drivers Than You or I," *Technology Review*, October 25, 2013, http://www.technologyreview.com/news/520746/data-shows-googles-robot-cars-are-smoother-safer-drivers-than-you-or-i/.

10. See ibid. for Chris Urmson's comments.

11. "The Self-Driving Car Logs More Miles on New Wheels" (Google corporate blog), August 7, 2012, http://googleblog.blogspot.co.uk/2012/08/the-self-driving-car-logs-more-miles-on.html.

12. As quoted in Heather Kelly, "Driverless Car Tech Gets Serious at CES," *CNN*, January 9, 2014, http://www.cnn.com/2014/01/09/tech/innovation/self-driving-cars-ces/.

13. For US accident statistics, see http://www.census.gov/compendia/statab/2012/tables/12s1103.pdf; for global accident statistics, see http://www.who.int/gho/road_safety/mortality/en/.

14. Information on collision avoidance systems can be found at http://www.iihs.org/iihs/topics/t/crash-avoidance-technologies/qanda.

15. As quoted in Burkhard Bilger, "Auto Correct: Has the Self-Driving Car at Last Arrived?," *New Yorker*, November 25, 2013, http://www.newyorker.com/reporting/2013/11/25/131125fa_fact_bilger?currentPage=all.

16. John Markoff, "Google's Next Phase in Driverless Cars: No Steering Wheel or Brake Pedals," *New York Times*, May 27, 2014, http://www.nytimes

.com/2014/05/28/technology/googles-next-phase-in-driverless-cars-no-brakes-or-steering-wheel.html.

17. Kevin Drum, "Driverless Cars Will Change Our Lives. Soon.," *Mother Jones* (blog), January 24, 2013, http://www.motherjones.com/kevin-drum/2013/01/driverless-cars-will-change-our-lives-soon.

18. Lila Shapiro, "Car Wash Workers Unionize in Los Angeles," *Huffington Post,* February 23, 2012, http://www.huffingtonpost.com/2012/02/23/car-wash-workers-unionize_n_1296060.html.

19. David Von Drehle, "The Robot Economy," *Time,* September 9, 2013, pp. 44–45.

20. Andrew Harris, "Chicago Cabbies Sue Over Unregulated Uber, Lyft Services," *Bloomberg News,* February 6, 2014, http://www.bloomberg.com/news/2014–02–06/chicago-cabbies-sue-over-unregulated-uber-lyft-services.html.

Chapter 8

1. For statistics on consumer spending, see Nelson D. Schwartz, "The Middle Class Is Steadily Eroding. Just Ask the Business World," *New York Times,* February 2, 2014, http://www.nytimes.com/2014/02/03/business/the-middle-class-is-steadily-eroding-just-ask-the-business-world.html.

2. Rob Cox and Eliza Rosenbaum, "The Beneficiaries of the Downturn," *New York Times,* December 28, 2008, http://www.nytimes.com/2008/12/29/business/29views.html. The famous "plutonomy memos" were also featured in Michael Moore's 2009 documentary film *Capitalism: A Love Story.*

3. Barry Z. Cynamon and Steven M. Fazzari, "Inequality, the Great Recession, and Slow Recovery," January 23, 2014, http://pages.wustl.edu/files/pages/imce/fazz/cyn-fazz_consinequ_130113.pdf.

4. Ibid.

5. Ibid., p. 18.

6. Mariacristina De Nardi, Eric French, and David Benson, "Consumption and the Great Recession," National Bureau of Economic Research, NBER Working Paper No. 17688, issued in December 2011, http://www.nber.org/papers/w17688.pdf.

7. Cynamon and Fazzari, "Inequality, the Great Recession, and Slow Recovery," p. 29.

8. Derek Thompson, "ESPN President: Wage Stagnation, Not Technology, Is the Biggest Threat to the TV Business," *The Atlantic,* August 22, 2013, http://www.theatlantic.com/business/archive/2013/08/espn-president-wage-stagnation-not-technology-is-the-biggest-threat-to-the-tv-business/278935/.

9. Jessica Hopper, "Waiting for Midnight, Hungry Families on Food Stamps Give Walmart 'Enormous Spike,'" *NBC News,* November 28, 2011, http://

rockcenter.nbcnews.com/_news/2011/11/28/9069519-waiting-for-midnight-hungry-families-on-food-stamps-give-walmart-enormous-spike.

10. Data Source: FRED, Federal Reserve Economic Data, Federal Reserve Bank of St. Louis: Corporate Profits After Tax (without IVA and CCAdj) [CP] and Retail Sales: Total (Excluding Food Services) [RSXFS]; http://research.stlouisfed.org/fred2/series/CP/; http://research.stlouisfed.org/fred2/series/RSXFS/; accessed April 29, 2014.

11. Joseph E. Stiglitz, "Inequality Is Holding Back the Recovery," *New York Times,* January 19, 2013, http://opinionator.blogs.nytimes.com/2013/01/19/inequality-is-holding-back-the-recovery.

12. Washington Center for Equitable Growth, interview with Robert Solow, January 14, 2014, video available at http://equitablegrowth.org/2014/01/14/1472/our-bob-solow-equitable-growth-interview-tuesday-focus-january-14–2014.

13. Paul Krugman, "Inequality and Recovery," *New York Times* (The Conscience of a Liberal blog), January 20, 2013, http://krugman.blogs.nytimes.com/2013/01/20/inequality-and-recovery/.

14. See, for example, Krugman's "Cogan, Taylor, and the Confidence Fairy," *New York Times* (The Conscience of a Liberal blog), March 19, 2013, http://krugman.blogs.nytimes.com/2013/03/19/cogan-taylor-and-the-confidence-fairy/.

15. Paul Krugman, "How Did Economists Get It So Wrong?" *New York Times Magazine,* September 2, 2009, http://www.nytimes.com/2009/09/06/magazine/06Economic-t.html.

16. John Maynard Keynes, *The General Theory of Employment, Interest and Money* (London: Macmillan, 1936), ch. 21, available online at http://gutenberg.net.au/ebooks03/0300071h/chap21.html.

17. For figures on US productivity, see US Bureau of Labor Statistics, Economic News Release, March 6, 2014, http://www.bls.gov/news.release/prod2.nr0.htm.

18. Lawrence Mishel, "Declining Value of the Federal Minimum Wage Is a Major Factor Driving Inequality," Economic Policy Institute, February 21, 2013, http://www.epi.org/publication/declining-federal-minimum-wage-inequality/.

19. Eric Schlosser, *Fast Food Nation: The Dark Side of the All-American Meal* (New York: Harper, 2004), p. 66.

20. Emmanuel Saez, "Striking It Richer: The Evolution of Top Incomes in the United States," University of California, Berkeley, September 3, 2013, http://elsa.berkeley.edu/~saez/saez-UStopincomes-2012.pdf.

21. Andrew G. Berg and Jonathan D. Ostry, "Inequality and Unsustainable Growth: Two Sides of the Same Coin?," International Monetary Fund, April 8, 2011, http://www.imf.org/external/pubs/ft/sdn/2011/sdn1108.pdf.

22. Andrew G. Berg and Jonathan D. Ostry, "Warning! Inequality May Be Hazardous to Your Growth," *iMFdirect,* April 8, 2011, http://blog-imfdirect.imf.org/2011/04/08/inequality-and-growth/.

23. Ellen Florian Kratz, "The Risk in Subprime," *CNN Money,* March 1, 2007, http://money.cnn.com/2007/02/28/magazines/fortune/subprime.fortune/index.htm?postversion=2007030117.

24. Senior Supervisors Group, "Progress Report on Counterparty Data," January 15, 2014, http://www.newyorkfed.org/newsevents/news/banking/2014/SSG_Progress_Report_on_Counterparty_January2014.pdf.

25. Noah Smith, "Drones Will Cause an Upheaval of Society Like We Haven't Seen in 700 Years," *Quartz,* March 11, 2014, http://qz.com/185945/drones-are-about-to-upheave-society-in-a-way-we-havent-seen-in-700-years.

26. Barry Bluestone and Mark Melnik, "After the Recovery: Help Needed," *Civic Ventures,* 2010, http://www.encore.org/files/research/JobsBluestonePaper3–5-10.pdf.

27. Andy Sharp and Masaaki Iwamoto, "Japan Real Wages Fall to Global Recession Low in Abe [Japanese Prime Minister] Risk," *Bloomberg Businessweek,* February 5, 2014, http://www.businessweek.com/news/2014–02–05/japan-real-wages-fall-to-global-recession-low-in-spending-risk.

28. On youth unemployment, see Ian Sivera, "Italy's Youth Unemployment at 42% as Jobless Rate Hits 37-Year High," *International Business Times,* January 8, 2014, http://www.ibtimes.co.uk/italys-jobless-rate-hits-37-year-record-high-youth-unemployment-reaches-41–6-1431445, and Ian Sivera, "Spain's Youth Unemployment Rate Hits 57.7% as Europe Faces a 'Lost Generation,'" *International Business Times,* January 8, 2014, http://www.ibtimes.co.uk/spains-youth-unemployment-rate-hits-57–7-europe-faces-lost-generation-1431480.

29. James M. Poterba, "Retirement Security in an Aging Society," National Bureau of Economic Research, NBER Working Paper No. 19930, issued in February 2014, http://www.nber.org/papers/w19930 and also http://www.nber.org/papers/w19930.pdf. See Table 9, p. 21.

30. Ibid., based on Table 15, p. 39; see the row labeled "Joint & Survivor, Male 65 and Female 60, 100% Survivor Income-Life Annuity." An alternate plan with a 3 percent annual increase starts at just $3,700 (or about $300 per month).

31. Carl Benedikt Frey and Michael A. Osborne, "The Future of Employment: How Susceptible Are Jobs to Computerisation?," Oxford Martin School, Programme on the Impacts of Future Technology, September 17, 2013, p. 38, http://www.futuretech.ox.ac.uk/sites/futuretech.ox.ac.uk/files/The_Future_of_Employment_OMS_Working_Paper_1.pdf.

32. For China population figures, see Deirdre Wang Morris, "China's Aging Population Threatens Its Manufacturing Might," CNBC, October 24, 2012, http://www.cnbc.com/id/49498720 and "World Population Ageing 2013," United Nations, Department of Economic and Social Affairs, Population Division, p. 32, http://www.un.org/en/development/desa/population/publications/pdf/ageing/WorldPopulationAgeing2013.pdf.

33. On the Chinese savings rate (which, as noted, is as high as 40 percent), see Keith B. Richburg, "Getting Chinese to Stop Saving and Start Spending Is a Hard Sell," *Washington Post,* July 5, 2012, http://www.washingtonpost.com/world/asia_pacific/getting-chinese-to-stop-saving-and-start-spending-is-a-hard-sell/2012/07/04/gJQAc7P6OW_story_1.html, and "China's Savings Rate World's Highest," *China People's Daily,* November 30, 2012, http://english.people.com.cn/90778/8040481.html.

34. Mike Riddell, "China's Investment/GDP Ratio Soars to a Totally Unsustainable 54.4%. Be Afraid," *Bond Vigilantes,* January 14, 2014, http://www.bondvigilantes.com/blog/2014/01/24/chinas-investmentgdp-ratio-soars-to-a-totally-unsustainable-54–4-be-afraid/.

35. Dexter Robert, "Expect China Deposit Rate Liberalization Within Two Years, Says Central Bank Head," *Bloomberg Businessweek,* March 11, 2014, http://www.businessweek.com/articles/2014–03–11/china-deposit-rate-liberalization-within-two-years-says-head-of-chinas-central-bank.

36. Shang-Jin Wei and Xiaobo Zhang, "Sex Ratios and Savings Rates: Evidence from 'Excess Men' in China," February 16, 2009, http://igov.berkeley.edu/sites/default/files/Shang-Jin.pdf.

37. Caroline Baum, "So Who's Stealing China's Manufacturing Jobs?," *Bloomberg News,* October 14, 2003, http://www.bloomberg.com/apps/news?pid=newsarchive&sid=aRI4bAft7Xw4.

38. On investment and the business cycle, see Paul Krugman, "Shocking Barro," *New York Times* (The Conscience of a Liberal blog), September 12, 2011, http://krugman.blogs.nytimes.com/2011/09/12/shocking-barro/.

Chapter 9

1. Stephen Hawking, Stuart Russell, Max Tegmark, and Frank Wilczek, "Stephen Hawking: 'Transcendence Looks at the Implications of Artificial Intelligence—But Are We Taking AI Seriously Enough?,'" *The Independent,* May 1, 2014, http://www.independent.co.uk/news/science/stephen-hawking-transcendence-looks-at-the-implications-of-artificial-intelligence-but-are-we-taking-ai-seriously-enough-9313474.html.

2. James Barrat, *Our Final Invention: Artificial Intelligence and the End of the Human Era* (New York: Thomas Dunne, 2013), pp. 196–197.

3. Yann LeCun, Google+ Post, October 28, 2013, https://plus.google.com/+YannLeCunPhD/posts/Qwj9EEkUJXY.

4. Gary Marcus, "Hyping Artificial Intelligence, Yet Again," *New Yorker* (Elements blog), January 1, 2014, http://www.newyorker.com/online/blogs/elements/2014/01/the-new-york-times-artificial-intelligence-hype-machine.html.

5. Vernor Vinge, "The Coming Technological Singularity: How to Survive in the Post-Human Era," NASA VISION-21 Symposium, March 30–31, 1993.

6. Ibid.

7. Robert M. Geraci, "The Cult of Kurzweil: Will Robots Save Our Souls?," *USC Religion Dispatches,* http://www.religiondispatches.org/archive/culture/4456/the_cult_of_kurzweil%3A_will_robots_save_our_souls/.

8. "Noam Chomsky: The Singularity Is Science Fiction!" (interview), YouTube, October 4, 2013, https://www.youtube.com/watch?v=0kICLG4Zg8s#t=1393.

9. As quoted in *IEEE Spectrum,* "Tech Luminaries Address Singularity," http://spectrum.ieee.org/computing/hardware/tech-luminaries-address-singularity.

10. Ibid.

11. James Hamblin, "But What Would the End of Humanity Mean for Me?," *The Atlantic,* May 9, 2014, http://www.theatlantic.com/health/archive/2014/05/but-what-does-the-end-of-humanity-mean-for-me/361931/.

12. Gary Marcus, "Why We Should Think About the Threat of Artificial Intelligence," *New Yorker* (Elements blog), October 24, 2013, http://www.newyorker.com/online/blogs/elements/2013/10/why-we-should-think-about-the-threat-of-artificial-intelligence.html.

13. P. Z. Myers, "Ray Kurzweil Does Not Understand the Brain," *Pharyngula Science Blog,* August 17, 2010, http://scienceblogs.com/pharyngula/2010/08/17/ray-kurzweil-does-not-understa/.

14. Barrat, *Our Final Invention: Artificial Intelligence and the End of the Human Era,* pp. 7–21.

15. Richard Feynman, "There's Plenty of Room at the Bottom," lecture at CalTech, December 29, 1959, full text available at http://www.zyvex.com/nanotech/feynman.html.

16. On federal research funding for nanotechnology, see John F. Sargent Jr., "The National Nanotechnology Initiative: Overview, Reauthorization, and Appropriations Issues," Congressional Research Service, December 17, 2013, https://www.fas.org/sgp/crs/misc/RL34401.pdf.

17. K. Eric Drexler, *Radical Abundance: How a Revolution in Nanotechnology Will Change Civilization* (New York: PublicAffairs, 2013), p. 205.

18. Ibid.

19. K. Eric Drexler, *Engines of Creation: The Coming Era of Nanotechnology* (New York: Anchor Books, 1986, 1990), p. 173.

20. Bill Joy, "Why the Future Doesn't Need Us," *Wired,* April 2000, http://www.wired.com/wired/archive/8.04/joy.html.

21. "Nanotechnology: Drexler and Smalley Make the Case For and Against 'Molecular Assemblers,'" *Chemical and Engineering News,* December 1, 2003, http://pubs.acs.org/cen/coverstory/8148/8148counterpoint.html.

22. Institute of Nanotechnology website, http://www.nano.org.uk/nano/nanotubes.php.

23. Luciana Gravotta, "Cheap Nanotech Filter Clears Hazardous Microbes and Chemicals from Drinking Water," *Scientific American,* May 7, 2013, http://www.scientificamerican.com/article/cheap-nanotech-filter-water/.

24. Drexler, *Radical Abundance,* pp. 147–148.

25. Ibid., p. 210.

Chapter 10

1. Interview between JFK and Walter Cronkite, September 2, 1963, https://www.youtube.com/watch?v=RsplVYbB7b8 8:00. Kennedy's comments on unemployment begin at about the 8-minute mark in this YouTube video.

2. "Skill Mismatch in Europe," European Centre for the Development of Vocational Training, June 2010, http://www.cedefop.europa.eu/EN/Files/9023_en.pdf?_ga=1.174939682.1636948377.1400554111.

3. Jock Finlayson, "The Plight of the Overeducated Worker," *Troy Media,* January 13, 2014, http://www.troymedia.com/2014/01/13/the-plight-of-the-overeducated-worker/.

4. Jin Zhu, "More Workers Say They Are Over-Educated," *China Daily,* February 8, 2013, http://europe.chinadaily.com.cn/china/2013–02/08/content_16213715.htm.

5. Hal Salzman, Daniel Kuehn, and B. Lindsay Lowell, "Guestworkers in the High-Skill U.S. Labor Market," Economic Policy Institute, April 24, 2013, http://www.epi.org/publication/bp359-guestworkers-high-skill-labor-market-analysis/.

6. Steven Brint, "The Educational Lottery," *Los Angeles Review of Books,* November 15, 2011, http://lareviewofbooks.org/essay/the-educational-lottery.

7. Nicholas Carr, "Transparency Through Opacity" (blog), *Rough Type,* May 5, 2014, http://www.roughtype.com/?p=4496.

8. Erik Brynjolfsson, "Race Against the Machine," presentation to the President's Council of Advisors on Science and Technology (PCAST), May 3, 2013, http://www.whitehouse.gov/sites/default/files/microsites/ostp/PCAST/PCAST_May3_Erik%20Brynjolfsson.pdf, p. 28.

9. Claire Cain Miller and Chi Birmingham, "A Vision of the Future from Those Likely to Invent It," *New York Times* (The Upshot), May 2, 2014, http://www.nytimes.com/interactive/2014/05/02/upshot/FUTURE.html.

10. F. A. Hayek, *Law, Legislation and Liberty, Volume 3: The Political Order of a Free People* (Chicago: University of Chicago Press, 1979), pp. 54–55.

11. Ibid., p. 55.

12. John Schmitt, Kris Warner, and Sarika Gupta, "The High Budgetary Cost of Incarceration," Center for Economic and Policy Research, June 2010, http://www.cepr.net/documents/publications/incarceration-2010–06.pdf.

13. John G. Fernald and Charles I. Jones, "The Future of US Economic Growth," *American Economic Review: Papers & Proceedings* 104, no. 5 (2014): 44–49, http://www.aeaweb.org/articles.php?doi=10.1257/aer.104.5.44.

14. Sam Peltzman, "The Effects of Automobile Safety Regulation," *Journal of Political Economy* 83, no. 4 (August 1975), http://www.jstor.org/discover/10.2307/1830396?uid=3739560&uid=2&uid=4&uid=3739256&sid=21103816422091.

15. Hanna Rosin, "The Overprotected Kid," *The Atlantic,* March 19, 2014, http://www.theatlantic.com/features/archive/2014/03/hey-parents-leave-those-kids-alone/358631/.

16. "Improving Social Security in Canada, Guaranteed Annual Income: A Supplementary Paper," Government of Canada, 1994, http://www.canadiansocialresearch.net/ssrgai.htm.

17. One analysis of costs and potential offsets from programs that could be eliminated is here: Danny Vinik, "Giving All Americans a Basic Income Would End Poverty," *Slate,* November 17, 2013, http://www.slate.com/blogs/business_insider/2013/11/17/american_basic_income_an_end_to_poverty.html.

18. Noah Smith, "The End of Labor: How to Protect Workers from the Rise of Robots," *The Atlantic,* January 14, 2013, http://www.theatlantic.com/business/archive/2013/01/the-end-of-labor-how-to-protect-workers-from-the-rise-of-robots/267135/.

19. Nelson D. Schwartz, "217,000 Jobs Added, Nudging Payrolls to Levels Before the Crisis," *New York Times,* June 6, 2014, http://www.nytimes.com/2014/06/07/business/labor-department-releases-jobs-data-for-may.html.

Conclusion

1. Shawn Sprague, "What Can Labor Productivity Tell Us About the U.S. Economy?," US Bureau of Labor Statistics, *Beyond the Numbers* 3, no. 12 (May 2014), http://www.bls.gov/opub/btn/volume-3/pdf/what-can-labor-productivity-tell-us-about-the-us-economy.pdf.

2. National Climate Assessment, "Welcome to the National Climate Assessment," *Global Change.gov,* n.d., http://nca2014.globalchange.gov/.

3. Stephen Lacey, "Chart: 2/3rds of Global Solar PV Has Been Installed in the Last 2.5 Years," *GreenTechMedia.com,* August 13, 2013, http://www.greentechmedia.com/articles/read/chart-2–3rds-of-global-solar-pv-has-been-connected-in-the-last-2.5-years.

4. Lauren Feeney, "Climate Change No Problem, Says Futurist Ray Kurzweil," *The Guardian,* February 21, 2011, http://www.theguardian.com/environment/2011/feb/21/ray-kurzweill-climate-change.

5. "Climate Change in the American Mind: Americans' Global Warming Beliefs and Attitudes in April 2013," Yale Project on Climate Change

Communication/George Mason University Center for Climate Change Communication, http://environment.yale.edu/climate-communication/files/Climate-Beliefs-April-2013.pdf.

6. Rebecca Riffkin, "Climate Change Not a Top Worry in U.S.," *Gallup Politics*, March 12, 2014, http://www.gallup.com/poll/167843/climate-change-not-top-worry.aspx.

INDEX

Martin Ford, the founder of a Silicon Valley–based software development firm, has over twenty-five years of experience in computer design and software development. The author of *The Lights in the Tunnel: Automation, Accelerating Technology, and the Economy of the Future*, he lives in Sunnyvale, California.

@MFordFuture

Photograph by Xiaoxiao Zhao